Advances in Intelligent Systems and Computing

Volume 996

The series "Advances in Intelligent Systems and Computing" contains publications on theory, applications, and design methods of Intelligent Systems and Intelligent Computing. Virtually all disciplines such as engineering, natural sciences, computer and information science, ICT, economics, business, e-commerce, environment, healthcare, life science are covered. The list of topics spans all the areas of modern intelligent systems and computing such as: computational intelligence, soft computing including neural networks, fuzzy systems, evolutionary computing and the fusion of these paradigms, social intelligence, ambient intelligence, computational neuroscience, artificial life, virtual worlds and society, cognitive science and systems, Perception and Vision, DNA and immune based systems, self-organizing and adaptive systems, e-Learning and teaching, human-centered and human-centric computing, recommender systems, intelligent control, robotics and mechatronics including human-machine teaming, knowledge-based paradigms, learning paradigms, machine ethics, intelligent data analysis, knowledge management, intelligent agents, intelligent decision making and support, intelligent network security, trust management, interactive entertainment, Web intelligence and multimedia.

The publications within "Advances in Intelligent Systems and Computing" are primarily proceedings of important conferences, symposia and congresses. They cover significant recent developments in the field, both of a foundational and applicable character. An important characteristic feature of the series is the short publication time and world-wide distribution. This permits a rapid and broad dissemination of research results.

** Indexing: The books of this series are submitted to ISI Proceedings, EI-Compendex, DBLP, SCOPUS, Google Scholar and Springerlink **

More information about this series at http://www.springer.com/series/11156

Rituparna Chaki · Agostino Cortesi ·
Khalid Saeed · Nabendu Chaki
Editors

Advanced Computing and Systems for Security

Volume Ten

 Springer

Editors
Rituparna Chaki
A.K. Choudhury School of Information
Technology
University of Calcutta
Kolkata, West Bengal, India

Khalid Saeed
Faculty of Computer Science
Bialystok University of Technology
Bialystok, Poland

Agostino Cortesi
Computer Science, DAIS
Università Ca' Foscari
Mestre, Venice, Italy

Nabendu Chaki
Department of Computer Science
and Engineering
University of Calcutta
Kolkata, West Bengal, India

ISSN 2194-5357   ISSN 2194-5365 (electronic)
Advances in Intelligent Systems and Computing
ISBN 978-981-13-8968-9   ISBN 978-981-13-8969-6 (eBook)
https://doi.org/10.1007/978-981-13-8969-6

This Springer imprint is published by the registered company Springer Nature Singapore Pte Ltd.
The registered company address is: 152 Beach Road, #21-01/04 Gateway East, Singapore 189721, Singapore

Preface

This volume contains the revised and improved version of papers presented at the 6th International Doctoral Symposium on Applied Computation and Security Systems (ACSS 2019) which took place in Kolkata, India, during 12–13 March 2019. The University of Calcutta in collaboration with Ca' Foscari University of Venice, Italy, and Bialystok University of Technology, Poland, organized the symposium. This symposium is unique in its characteristic of providing Ph.D. scholars with an opportunity to share the preliminary results of their work in an international context and be actively supported towards their first publication in a scientific volume.

On our pursuit of continuous excellence, we aim to include the emergent research domains in the scope of the symposium each year. This helps ACSS to stay in tune with the evolving research trends. The sixth year of the symposium was marked with a significant improvement in overall quality of papers, besides some very interesting papers in the domain of security and software engineering. We are grateful to the Programme Committee members for sharing their expertise and taking time off from their busy schedule to complete the review of the papers with utmost sincerity. The reviewers have pointed out the improvement areas for each paper they reviewed—and we believe that these suggestions would go a long way in improving the overall quality of research among the scholars. We have invited eminent researchers from academia and industry to chair the sessions which matched their research interests. As in previous years, the session chairs for each session had a prior go-through of each paper to be presented during the respective sessions. This is done to make it more interesting as we found deep involvement of the session chairs in mentoring the young scholars during their presentations.

The evolution of ACSS is an interesting process. We have noticed the emergence of security as a very important aspect of research, due to the overwhelming presence of IoT in every aspect of life.

The indexing initiatives from Springer have drawn a large number of high-quality submissions from scholars in India and abroad. ACSS continues with the tradition of double-blind review process by the PC members and by external reviewers. The reviewers mainly considered the technical aspect and novelty of each paper, besides the validation of each work. This being a doctoral symposium, clarity of presentation was also given importance.

The Technical Program Committee for the symposium selected only 18 papers for publication out of 42 submissions.

We would like to take this opportunity to thank all the members of the Programme Committee and the external reviewers for their excellent and time-bound review works.

We thank the members of the Organizing Committee, whose sincere efforts before and during the symposium have resulted in a friendly and engaging event, where the discussions and suggestions during and after the paper presentations create a sense of community that is so important for supporting the growth of young researchers.

We thank Springer for sponsoring the best paper award. We would also like to thank ACM for the continuous support towards the success of the symposium. We appreciate the initiative and support from Mr. Aninda Bose and his colleagues in Springer Nature for their strong support towards publishing this post-symposium book in the series "Advances in Intelligent Systems and Computing". Last but not least, we thank all the authors without whom the symposium would not have reached up to this standard.

On behalf of the editorial team of ACSS 2019, we sincerely hope that ACSS 2019 and the works discussed in the symposium will be beneficial to all its readers and motivate them towards even better works.

Kolkata, India   Rituparna Chaki
Mestre, Venice, Italy   Agostino Cortesi
Bialystok, Poland   Khalid Saeed
Kolkata, India   Nabendu Chaki

Contents

High Performance Computing

About the Editors

Rituparna Chaki is Professor of Information Technology in the University of Calcutta, India. She received her Ph.D. Degree from Jadavpur University in India in 2003. Before this she completed B.Tech. and M.Tech. in Computer Science & Engineering from the University of Calcutta in 1995 and 1997 respectively. She has served as a System Executive in the Ministry of Steel, Government of India for nine years, before joining the academics in 2005 as a Reader of Computer Science & Engineering in the West Bengal University of Technology, India. She is with the University of Calcutta since 2013. Her area of research includes Optical networks, Sensor networks, Mobile ad hoc networks, Internet of Things, Data Mining, etc. She has nearly 100 publications to her credit. Dr. Chaki has also served in the program committees of different international conferences. She has been a regular Visiting Professor in the AGH University of Science & Technology, Poland for the last few years. Dr. Chaki has co-authored a couple of books published by CRC Press, USA.

Agostino Cortesi, Ph.D., is a Full Professor of Computer Science at Ca' Foscari University, Venice, Italy. He served as Dean of the Computer Science studies, as Department Chair, and as Vice-Rector for quality assessment and institutional affairs. His main research interests concern programming languages theory, software engineering, and static analysis techniques, with particular emphasis on security applications. He published more than 110 papers in high-level international journals and proceedings of international conferences. His h-index is 16 according to Scopus, and 24 according to Google Scholar. Agostino served several times as member (or chair) of program committees of international conferences (e.g., SAS, VMCAI, CSF, CISIM, ACM SAC) and he is in the editorial boards of the journals "Computer Languages, Systems and Structures" and "Journal of Universal Computer Science". Currently, he holds the chairs of "Software Engineering", "Program Analysis and Verification", "Computer Networks and Information Systems" and "Data Programming".

Khalid Saeed is a Full Professor in the Faculty of Computer Science, Bialystok University of Technology, Bialystok, Poland. He received the B.Sc. Degree in Electrical and Electronics Engineering in 1976 from Baghdad University in 1976, the M.Sc. and Ph.D. Degrees from Wroclaw University of Technology, in Poland in 1978 and 1981, respectively. He received his D.Sc. Degree (Habilitation) in Computer Science from the Polish Academy of Sciences in Warsaw in 2007. He was a Visiting Professor of Computer Science with the Bialystok University of Technology, where he is now working as a Full Professor He was with AGH University of Science and Technology in the years 2008–2014. He is also working as a Professor with the Faculty of Mathematics and Information Sciences at Warsaw University of Technology. His areas of interest are Biometrics, Image Analysis and Processing and Computer Information Systems. He has published more than 220 publications, edited 28 books, Journals and Conference Proceedings, 10 text and reference books. He supervised more than 130 M.Sc. and 16 Ph.D. theses. He gave more than 40 invited lectures and keynotes in different conferences and universities in Europe, China, India, South Korea and Japan on Biometrics, Image Analysis and Processing. He received more than 20 academic awards. Khalid Saeed is a member of more than 20 editorial boards of international journals and conferences. He is an IEEE Senior Member and has been selected as IEEE Distinguished Speaker for 2011–2016. Khalid Saeed is the Editor-in-Chief of International Journal of Biometrics with Inderscience Publishers.

Nabendu Chaki is a Professor in the Department of Computer Science & Engineering, University of Calcutta, Kolkata, India. Dr. Chaki did his first graduation in Physics from the legendary Presidency College in Kolkata and then in Computer Science & Engineering from the University of Calcutta. He has completed Ph.D. in 2000 from Jadavpur University, India. He is sharing 6 international patents including 4 U.S. patents with his students. Prof. Chaki has been quite active in developing international standards for Software Engineering and Cloud Computing as a member of Global Directory (GD) for ISO-IEC. Besides editing more than 25 book volumes, Nabendu has authored 6 text and research books and has more than 150 Scopus Indexed research papers in Journals and International conferences. His areas of research interests include distributed systems, image processing and software engineering. Dr. Chaki has served as a Research Faculty in the Ph.D. program in Software Engineering in U.S. Naval Postgraduate School, Monterey, CA. He is a visiting faculty member for many Universities in India and abroad. Besides being in the editorial board for several international journals, he has also served in the committees of over 50 international conferences. Prof. Chaki is the founder Chair of ACM Professional Chapter in Kolkata.

Security Systems

A Lightweight Security Protocol for IoT Using Merkle Hash Tree and Chaotic Cryptography

Nashreen Nesa⊚ **and Indrajit Banerjee**

Abstract Security is one of the primary concerns in an Internet of things (IoT) environment as they are deployed in critical applications that directly affect human lives. For this purpose, a security protocol that involves both authentication of deployed IoT devices and encryption of generated data is proposed in this paper. Encryption algorithms based on chaos are known to satisfy the basic requirements of the cryptosystem such as high sensitivity, high computational speed and high security. The chaos-based encryption algorithm is built upon a modified quadratic map named as quadratic sinusoidal map which exhibits better array of chaotic regime when compared to the traditional logistic map. The authentication protocol, on the other hand, is based on Merkle hash tree that has been improved to adapt to an IoT environment. The control parameters and the initial condition for the map are derived from the Merkle hash tree. The proposed algorithm involves cryptographic operations that incur very low computational cost and requires relatively little storage and at the same time are highly resilient to security attacks.

Keywords Chaos theory · Merkle hash tree · IoT · Security · Encryption

1 Introduction

Advances in the field of Internet technologies, wireless sensor networks (WSN) and other wireless technologies have resulted in the new wave of Internet of things (IoT) services which have completely changed the lifestyle of the society. However, these services have also opened new possibilities of attacks and other fraudulent

N. Nesa (✉) · I. Banerjee
Department of Information Technology, Indian Institute of Engineering
Science and Technology, Shibpur, Howrah 711103, West Bengal, India
e-mail: nashreennesa.rs2016@it.iiests.ac.in

I. Banerjee
e-mail: ibanerjee@it.iiests.ac.in
URL: http://www.iiests.ac.in

© Springer Nature Singapore Pte Ltd. 2020
R. Chaki et al. (eds.), *Advanced Computing and Systems for Security*,
Advances in Intelligent Systems and Computing 996,
https://doi.org/10.1007/978-981-13-8969-6_1

activities as it has now become convenient for attackers to get unauthorized remote access to data anytime anywhere [1]. Every smart object in an IoT environment can become the entry point of malicious activity. Therefore, it is essential for every communicating device in the network to verify its identity so that no unauthorized device can take part in communication [2]. Security algorithms for IoT applications must ensure key requirements such as source authentication, confidentiality, data integrity and resistance against attacks [3] among other needs. Thus, in our work, we have proposed a security protocol for IoT that secures the data generated from the sensing devices in the form of encryption. Furthermore, for authenticating the devices, we have modified the Merkle hash tree (MHT) as the authentication protocol. MHT has been widely used as an authentication protocol [4, 5] and integrity solution [6] in the available literature. Since IoT devices have unique 96-bit binary numbers (24 hex digits) radio-frequency identification (RFID) tags for device identification, we have utilized it for the purpose of authentication by exploiting its uniqueness. The encryption algorithm in our work is based on chaotic cryptography. Chaos is a popular theory in numerous natural and laboratory systems, which is widely adopted in designing encryption algorithms [7, 8] or creation of pseudorandom numbers for encryption [9, 10]. In this paper, we have proposed a sinusoidal chaotic map based on the tradition quadratic map [11] that is used for encryption; the initial condition and the control parameters of which are produced from the MHT that forms the basis upon which the maps are created. The novelty of our work lies in the utilization of the RFID tags of the devices as an initial vector for the generation of the MHT. That is, for each leaf, the value ϕ is the hash value of the RFID of the IoT device. Our work uses lightweight computation modules, such as one-way hash functions and bitwise exclusive-or operation, for designing the secure data fusion protocol. Besides being a lightweight computation tool, the use of hash operations also preserves anonymity since the hash values are impossible to regenerate. Although researchers have investigated the concept of designing cryptosystem based on chaotic maps in the past [7, 12–14], but to the best of our knowledge, this is the first attempt at combining MHT with a novel chaotic map in order to achieve security in IoT applications.

The remainder of the paper is organized as follows: a related study on the recent trends in research is presented in Sect. 2, followed by the introduction of concepts of Merkle hash tree with its key definitions in Sect. 3. Details about our proposed modified sinusoidal quadratic map are presented in Sect. 4. Next, in Sect. 5, a description of all the phases in our proposed architecture is given. Section 6 describes an experimental scenario of our proposed algorithm for easy understanding, followed by the security analysis of the algorithms in Sect. 7. Finally, the paper is concluded with its ending remarks in Sect. 8.

2 Related Study

Although there exists a large volume of literature in the domain of network security, few of those work are focussed on the specific requirements of IoT. Although the research trend is growing, it will take few more years to catch up. Some of the recent works in the domain of IoT security are reviewed in this section. A survey reviewing the current research trends and open issues is presented in [15] where the authors discussed the current research works in authentication, encryption, trust management and secure routing classified according to the layer in the network stack. Another in-depth survey of IoT security from data perspective is presented in [16] where the authors integrated the concepts IoT architectures with the data life cycle. Additionally, the paper presents a three-dimensional approach to exploring IoT security grouped into one-stop, multistop and end-application dimensions. There is a need for designing lightweight protocols in order to suit the resource-constrained requirements for IoT objects. Keeping this in mind, the authors in [17] proposed a lightweight multicast authentication scheme for small-scale IoT applications that is inspired by the Nyberg's fast one-way accumulator. Detailed security and performance analysis against the seven performance aspects was carried out for the proposed scheme and found it to be superior in comparison with other works in the literature. Next, addressing the cyber-security requirements for the IoT, the authors in [18] presented a system architecture for Unit and Ubiquitous IoT system (U2IoT) that comprises three layers, i.e. the perception, network and application layers. The proposed scheme was analysed to be secure against data access, data sharing and authority transfer. The authors in [19] presented a IoT-based device management system based on IETF CoAP protocol. The proposed system which includes a low-power Wi-fi sensor device, a device management server and an IoT application is lightweight considering the specific requirements of IoT devices. A distributed IoT system architecture is presented in [20] where the authors proposed an authentication protocol to meet the security requirements of anonymity, traceability and resistance to various attacks. The proposed scheme promises low computational overhead rendering it suitable for resource-constrained systems such as IoT. Later, in [21], the authors proposed BSN-Care, a secure IoT-based healthcare system using body sensor network (BSN) that satisfies all the security requirements specific to healthcare. The proposed security solution is validated through extensive security analysis pertaining to mutual authentication, anonymity, secure localization and data security. The authors in [22] developed a security system for IoT devices to assess threats by categorizing them into three levels of intensity, i.e. "secure", "moderately secure" and "insecure". The assessment is based on the four security dimensions which are confidentiality, integrity, access control and resistance against reflexive attacks. Recently, with an attempt at integrating blockchain technology into IoT, the authors in [23] carried out a study to highlight the issues and challenges involved in the realization of the idea and resolved some of those issues related to it. In addition, several concepts of IoT-specific threats, impact of blockchain technology on IoT and their security and performance characteristics are discussed. In [24], the authors presented a secure IoT-based traffic controlling system that classifies traffic data into good or bad. The classification model used is SVM which has been implemented using Raspberry Pi and Scikit. The performance of the model was verified using a dataset that comprises of raw traffic data from three cities in London spanning a period of 5 years. In [25], the authors propose a methodology with an idea of certifying security for large-scale IoT deployments. The proposed mechanism, in particular, is beneficial for testers and consumers dedicated to developing and deploying of trusted IoT devices by providing them an opportunity to assess security solutions with their labels which are a direct outcome of the certification process.

It is evident that none of prior works in this domain collectively solves the scalability, authentication, encryption and resource-constraint issues of IoT systems. Therefore, in the light of these shortcomings, our proposed security protocol ensures authentication, encryption and scalability through the use of a novel approach that integrates Merkle hash tree and chaos theory. Our proposed system is lightweight and efficient and requires very little storage, making it an ideal choice for achieving security in IoT applications.

3 Merkle Hash Tree

Merkle hash tree (MHT) was first introduced in 1989 by Merkle [26] and has since then been used for verification and integrity checking by various applications. It is a popular technique among the Git and Bitcoin community for authenticating users. MHTs have mostly been used as authentication schemes [5, 27]. A typical MHT is a binary tree in which the nodes of the tree are simple hash values. The nodes at the lowest level (leaf nodes) could be an arbitrary hash values or the hash values generated from pseudorandom numbers, whereas the nodes at the intermediate levels are the hashes of their immediate children. The root of the MHT is unique since the collision resistance property of hash function ensures that no two hash values differing by atleast 1 bit should be same. To adapt to an IoT environment, the traditional definition of MHT concepts has been modified.

Definition 1 (*Merkle Hash assignment*) Let T be a MHT created in an IoT setup with n devices, i.e. having $\log_2 n+1$ levels, and let $RFID_i$ be the RFID of ith IoT device in T. The Merkle hash assignment associated with the device with $RFID_i$ of T at level 1, denoted as $\phi_{1,j}$ is computed as

$$\phi_{1,j} = H(RFID_j) \tag{1}$$

Similarly, the hash values of all node i except for the leaf nodes at level j denoted as $\phi_{i,j}$ are computed by the following function:

$$\phi_{i,j} = H(\phi_{i-1,2j-1}||\phi_{i-1,2j}) \tag{2}$$

where '$||$' denotes the concatenation operator and $H(.)$ is the hash function.

According to Definition 1, the Merkle hash value associated with a device D_i is the result of a hash function applied to its RFID tag. To ensure the integrity of the Merkle hash value of the trusted centre (TC), we assume that the root value ϕ_{root} is tamper-resistant whose credentials have been thoroughly checked by top-level security system. Each leaf node in the constructed MHT can be verified through its Merkle hash path θ which is defined next.

Fig. 1 Proposed sinusoidal chaotic map for $a = 4$

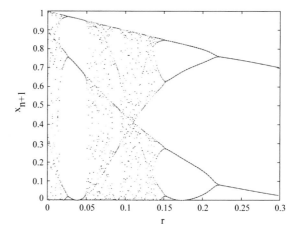

Definition 2 (*Merkle Hash Path*) For each level $l < \log_2 n + 1$ (height of the tree), we define Merkle hash path θ to be the ϕ values of all the sibling nodes at each level l on the path connecting the leaf to the root node. The Merkle hash path signifying the authentication data at level l is then the set $\theta_l | 1 \leq l \leq \log_2 n + 1$.

4 Modified Sinusoidal Quadratic Map

Chaotic maps are defined as mathematical functions that characterize the chaotic behaviour of the system and which depends on their initial conditions and control parameters. Our proposed chaotic map inspired by the classical quadratic map [11] is given as

$$x_{n+1} = 1 - \sin(r + ax_n^2) \text{ for } a > 3 \tag{3}$$

where the initial condition x_0, a and r are the control parameters. For our proposed equation, the value of a must be above 3, i.e. $a > 3$ in order to be chaotic. The bifurcation diagram of the proposed map is shown in Fig. 1 where the solution was iterated for different values of r for a particular value of $a(a = 4)$. Bifurcation diagrams in a dynamical system are characteristic plots that define the behaviour of a system when its control parameters undergo some change [28].

5 Proposed Security Protocol

Since our proposed architecture relates to an IoT scenario, adapting the known concepts of MHT and chaos theory to an IoT environment was necessary. The processes are explained in detail next.

Algorithm 1: Authentication Algorithm using MHT

Input: Number of devices in the network n
Output: Authenticated Result

```
/* Construction of Merkle tree                                    */
```
1 Number of levels $l \leftarrow \log_2 n + 1$
2 **foreach** *level* $j \in l$ **do**
3   **foreach** *device* $i \in N$ **do**
4   **if** $\phi_{i,j}$ *is a leaf node* **then**
5   Calculate $\phi_{i,j} \leftarrow H(RFID_i)$

6   **else**
7          $\phi_{i,j} \leftarrow H(\phi_{2i-1,j-1} || \phi_{2i,j-1})$

```
/* Authenticating an IoT device with RFID X                      */
```
8 **if** $\phi_{i,j} == \phi_{root}$ **then**
9   X is authenticated to be a valid device

10 **else**
11   A threat is detected

5.1 Registration

Registration is the first phase where all the n IoT devices $D_i(i = 1, \ldots, n)$ wish to form a network register themselves with the trusted centre (TC) with their designated n unique RFID tags. TC then constructs the tree by deriving the hash value of each $RFID_i$ and stores them in a table for future authentication.

5.2 Authenticating the Devices Using MHT

As mentioned earlier, authentication is performed by TC which is assumed to be secured from any form of attacks and whose credentials are verified from the top-level security system. Any device D_i wishing to initiate data transfer has to first test for its authenticity. This is done by sending a request message REQ_i to TC that indicates its desire for communication. TC, on receiving REQ_i, asks for the proof that D_i belongs to the network and is a valid device. D_i now sends hash values of the Merkle hash path (θ) for authentication. On receiving the proofs, TC calculates the hash value using Eq. 2, and checks if its stored hash of the root (i.e. h_{root}) is equal to the calculated hash. If the two hashes match, the device D_i is an authenticated device and can proceed to the process of data exchange. Algorithm 1 illustrates the process of authentication for our proposed system where each device presented as a leaf node is authenticated by recursively computing and concatenating the hash values along the Merkle hash path. Since only the hash functions are computed, the computation cost of verification is very low.

5.3   Establishment of Keys

In our work, the key of encryption algorithm is produced by our proposed chaotic map because of its marked nature of randomness. It is a known fact that the more random a key is, the more difficult it is for an attacker to break. Therefore, on the basis of the bifurcation diagram, it is easy to note the areas the map produces random chaotic behaviour. The control parameters needed to generate the bifurcation diagram comes from the MHT. Our proposed chaotic map takes as input three control parameters : a pre-shared session key $\mathcal{S}$, a key $\mathcal{K}$ which acts as the initial condition of the map and the number of iterations *Itr* that signifies the number of times $\mathcal{K}$ is iterated in the map. Based on the number of levels l, keys are produced. These keys are nothing but the value of ϕ at each node in the Merkle hash path θ for the device D_i, after which $\mathcal{K}$ is calculated as follows

$$\mathcal{K} = key_1 \oplus key_2 \oplus key_3 \oplus ... \oplus key_l$$

The value of $\mathcal{K}$ in binary is converted into decimal that serves as the initial condition in the chaotic map. The pre-shared session key $\mathcal{S}$ is generated which is a random number such that $\mathcal{S} > 3$. This is set keeping in mind that our proposed chaotic map is chaotic in this range. The iteration number *Itr* is calculated using both the values of $\mathcal{K}$ and $\mathcal{S}$. The number of digits in $\mathcal{K}$, say, *dig* is estimated. Now, in the generated value of $\mathcal{S}$, *dig* digits after decimal is extracted and summed up with the value of $\mathcal{K}$ that yields the iteration number.

Theorem 1 *The key space size of our proposed encryption algorithm is $2^{280} \times l$ where l denotes the number of levels in the Merkle hash tree.*

Proof Since the hash function used in our protocol is SHA-1, which produces a 160-bit binary output, the key space required for $\mathcal{K}$ alone is 2^{160}. Furthermore, in order to ensure the dynamical system falls in the chaotic regime, the range of *a* that is also the pre-shared session key is restricted to 32-digit values for $a > 3$. This value *a* or $\mathcal{S}$ is a 32-bit decimal number that is generated randomly. Since the ASCII table supported by MATLAB is composed of 128 values, each of these 40 hex digits (output of SHA-1) are mapped to its corresponding ASCII, the maximum value of which is 128. Therefore, the maximum value of the hash value at each node, i.e. key_1, key_2,...,key_l (l is the number of levels), cannot exceed $40 \times 128 = 5120$ which requires ≈ 13 bits each to represent. Therefore, the value of $\mathcal{K}$ which is the XOR operation of key_1, key_2, ..., key_l also comprises of 13 bits. Hence, the iteration number *Itr* that is dependent on the number of digits in $\mathcal{K}$ should also not exceed $13 + 1$ (for carry) bits. Summing it all up, the key size for our proposed algorithm is $(2^{160} \times l) \times 10^{32} \times 2^{13} \approx 2^{280} \times l$, where *l* is the number of levels in the Merkle hash tree.

Algorithm 2: Proposed Chaotic Encryption algorithm

Input: Raw data $\mathcal{P}$
Output: Encrypted data $\mathcal{C}$

1 Generate the keys $key_1, key_2, .., key_l$ based on the number of levels l
2 Final key $\mathcal{K} \leftarrow key_1 \oplus key_2 \oplus, ...key_n l$
3 Convert $\mathcal{K}$ into binary
4 $dig \leftarrow$ number of decimal digits of $\mathcal{K}$
5 $\delta \leftarrow$ take dig digits after decimal from pre-shared session key $\mathcal{S}$
6 Iteration $Itr \leftarrow \mathcal{K} + \delta$
  /* Set initial condition $\mathcal{K}$ and $\mathcal{S}$ as control parameter and
       iterate in the chaotic map Itr number of times   */
7 ChaosVal $\leftarrow$ Chaos($\mathcal{K}, \mathcal{S}, Itr$)
8 $\vartheta \leftarrow$ ChaosVal $\times Itr$
9 **if** $\mathcal{P}$ *is the first plaintext after registration* **then** $Cipher^t \leftarrow \mathcal{P}^t \oplus \vartheta$
10 **else** $Cipher^t \leftarrow \mathcal{P}^t \oplus \vartheta \oplus Cipher^{t-1}$
11 $\mathcal{C} \leftarrow C^t | \mathcal{N}^{(0)}$
12 **return** $\mathcal{C}$

5.4 Data Encryption/Decryption

The value of ϑ obtained after the iteration process of the chaotic map is then combined with the plain text $\mathcal{P}$ using a XOR operation along with the previous ciphertext value. The inclusion of previous ciphertext for XOR operation was adopted for ensuring dynamic feedback in our proposed architecture. Thus, at any instant t, the encrypted data will be given by $Cipher = \mathcal{P}^t \oplus \vartheta \oplus Cipher^{t-1}$. Algorithm 2 displays the essential steps of our chaos-based encryption/decryption algorithm. This function takes as an input a 256-bit plaintext $\mathcal{P}^t$ data. In order to add provision for integrity check, an alternative coding approach that appends a count of the '0' bits $N^{(0)}$ in C^t before communicating it to TC. The new message, $\mathcal{C}$, would be only be $\log_2$ 256-bit = 8 bits longer than the original 256-bit message, $Cipher^t$. After appending the zero count $N^{(0)}$, the final ciphertext $\mathcal{C}$ is sent to TC for data fusion through the communication medium. TC, on receiving $\mathcal{C}$, first checks whether the message has been tampered with by comparing the last 8 bits that signifies $N^{(0)}$ with the number of zeros in the first 256 bits in $\mathcal{C}$. If the values do not match the message, a security threat is detected and subsequent actions are undertaken to remedy the problem. If, however, the values match, the ciphertext is assumed to be free of any tampering by an intruder and thus is processed to extract the plaintext. In our work, since the Merkle hash values ϕ are used for modulation in the chaotic map, which is known to both D_i and TC, both can generate the chaotic initial value $\mathcal{K}$ and the Itr value individually. In the decryption process, utilizing the symmetric property of XOR operation TC decrypts the received data $Cipher^t$ as $Cipher^t \oplus \vartheta \oplus Cipher^{t-1} = (\mathcal{P}^t \oplus \vartheta \oplus Cipher^{t-1}) \oplus \vartheta \oplus Cipher^{t-1}$ which equals $\mathcal{P}^t$.

6 Experiment

For the sake of simplicity in our experiment, we have simulated an IoT environment consisting of 4 IoT devices D_1, D_2, D_3, D_4 in MATLAB, each generating time-varying data P^t every t time instant. For instance, sensory data (referred to as the plaintext) generated by device D_2 is given as $\mathcal{P}_2 = P_2^1, P_2^2, \ldots, P_2^{t-1}, P_2^t$. For our proposed chaotic map $x_{n+1} = 1 - \sin(r + ax_n^2)$ for $a > 3$, the control parameters are the pre-shared session key $\mathcal{S}$, the initial condition x_0 denoted as $\mathcal{K}$ and the iteration number Itr.

Registration All 4 IoT devices in the network register themselves with the trusted centre (TC) with their designated RFID tags, $RFID_i | i = 1, \ldots, 4$. RFIDs are 96-bit binary numbers or 24 hex digits. For our experiment, we have used random 24 hex numbers as RFIDs as shown in Table 1. TC creates a MHT with all the devices in the network, in which all the leaf nodes represent the RFIDs of the devices. The tree is constructed as shown in Fig. 2. In addition, TC maintains a table where RFID tag of each device is stored. A randomly generated pre-shared session key $\mathcal{S} > 3$ is stored for each time instant t. This session key is used by all the devices for communication for each session, at the expiration of which the session key $\mathcal{S}$ becomes obsolete.

Device authentication Referring to Fig. 2, suppose the device D_4 decides to initiate a communication and does so by sending a request message REQ_4 to TC. TC on receiving REQ_4 asks for the proof that D_4 belongs to the network and is a valid device. D_4 now sends hash values of the Merkle hash path for authentication. That is, D_4 sends the value of $\phi_{4,1}$, $\phi_{3,1}$ and $\phi_{1,2}$ as authentication proofs to the TC. The

Table 1 RFID tags corresponding to each device for our experiment

Device	RFID tags
D_1	45 3d 6c e1 48 16 85 57 e3 29 c5 89
D_2	77 c0 23 0e b5 0e 39 63 3a 48 5b bf
D_3	2e e0 62 6d 14 ca e6 83 18 7a e7 9e
D_4	ba 9d 08 f4 2b 4b 5e 23 51 d5 70 2a

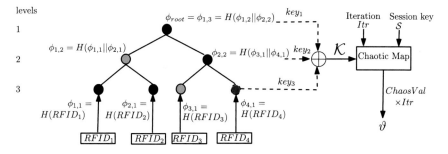

Fig. 2 Key generation using Merkle hash tree together with the proposed chaotic map

Table 2 Experimental Merkle hash values ϕ (40 digits each) of all nodes in the Merkle hash tree

Nodes	Merkle hash values
ϕ_{root}	d3e7feaed9d3fb1da6ea67597391c98f88a6dd94
$\phi_{1,2}$	faa426e4266d5b9f06751c60b1154ad707c79661
$\phi_{2,2}$	f330173cda222d69b62305b797a7ba1076e71b2b
$\phi_{1,1}$	9044022c3f4868497ec473c1ff540968e39f090c
$\phi_{2,1}$	dae4c09b389bf268ab79b51a56dbe430bb9eebda
$\phi_{3,1}$	dc1a5dc333b5a9afd8b9b28d89924e203f0b4b70
$\phi_{4,1}$	2dc408e9c8e7c3d95d03058bd59bf5627e8eee9a

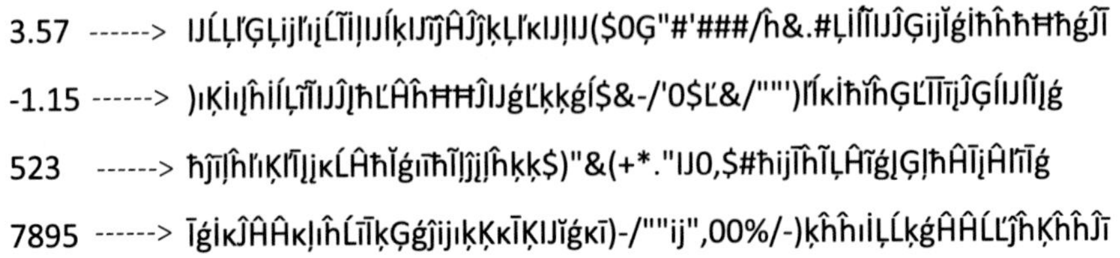

3.57 ------> ıJĹĻÍʿGĻijÍʿıĹTĩ̄Į|ıJĮ́kıJTĵ̃ĂĴ̃ĵ̨kĻ̨Í̀kıJ|ıJ($0G̃"#'###/ħ&.#Ļįĩ̄ıJĴ̃GįjĨǵiħħ̃ħ̃Ħħ̃ǵJ̄ı̄

-1.15 ------>)ıķiı|ħiĺĻ̄̃ĩ̄ıJĴ̃|ħĿ́Ĥ̃ħ̃ĦĦJ̃ıJǵĽ̨kķǵĨ́$&-/'0$Ŀ́&/""")Í̃kiħ̃ĩ̄ħ̃GĿ̄Tĩ̄J̃G̃ı̃JĨ́|ǵ

523 ------> ħ̃JĨ̄|ħ̃Íʿ̄kÍ̃T̨|ĵ̃kĹ́Ĥ̃ħ̃Ī́ǵıı̄ħ̃Ĩ̃|ĵ̃j|ħ̃kķ$)"&(+*."ıJO,$#ħijĨ̃ħĮ̃ĻʿĤ̃Īǵ|G̃|ħ̃Ĥ̃Į̄|Ĥ̃Ī̄ǵ

7895 ------> Ī́ǵikĴ̃Ĥ̃Ĥ̃ĸ|ıħ̃Ĺ̄Ī̄kĢ̃ǵĴ̃ijıķĶ̨kĶ̨kĮ̄KıJÍ̄ǵkĪ̄)-/""ij",00%/-)ķħ̃ħ̃iı̨ĻĿ̨kǵĤ̃Ĥ̃Ĺ̨Ĵ́ħ̃Ķ̃ħ̃ħ̃J̄ı̄

Fig. 3 Plaintext/sensor readings generating ciphertext for all 4 devices in our experiment

proofs/sibling nodes are highlighted in blue and the initiator node $\phi_{4,1}$ in red in Fig. 2 (best viewed in colour print). TC on receiving the proofs calculates the resultant hash value from the individual hash values received from D_4 according to Eq. 2, as

$$\phi_{1,3} = H(\phi_{1,2}||\phi_{2,2})$$
$$= H(\phi_{1,2}||H(\phi_{3,1}||\phi_{4,1}))$$
$$= \phi_{root}$$

TC now compares the resultant value $\phi_{1,3}$ with its own hash ϕ_{root} (Refer Table 2). The $\phi_{1,3}$ obtained in the previous step matches with the stored value of ϕ_{root} in our experiment, thereby authenticating the device D_4, and hence, it can proceed to the process of data exchange.

Key Generation and Exchange For the device D_4, the keys or ϕ values are presented in Table 2. After converting them to their ASCII and summing the resultant values, the final key $\mathcal{K}$ is the XOR of all 3 keys key_1, key_2 and key_3. In our experiment, value of $\mathcal{K}$ (note that the XOR operations are all done in binary, in order to save space, the values are replaced by their decimal equivalent) is

$$\mathcal{K} = 2751 \oplus 2567 \oplus 2907$$
$$= 184$$

The value of $\mathcal{K} = 184$ is the initial condition for the map generation. At this point, a shared session key $\mathcal{S}$ is generated between the TC and D_4. A random value $\mathcal{S} > 3$

is generated such that it falls in the chaotic region of the proposed map. For our experiment, we have randomly generated the value as

$$\mathcal{S} = 3.74012756841705806820641555553235$$

Next, the value of *Itr* is calculated by extracting the number of digits in $\mathcal{K}$ as *dig*. In our experiment, *dig* = 3. Adding *dig* number of digits after decimal of the value $\mathcal{S}$ to the value of $\mathcal{K}$ yields the value of *Itr*. That is,

$$Itr = 4 \text{ (value of dig) digits after decimal of } \mathcal{S} + \mathcal{K}$$
$$= 740 + 184 = 924$$

Encryption In our experiment, the values of *ChaosVal* and ϑ after the map is iterated a *Itr* number of times are as follows :

$$ChaosVal = 0.1525$$
$$\vartheta = ChaosVal \times Itr$$
$$= 0.1525 \times 924 = 140.91$$

The plaintext produced by D_4 at time instant t and (t-1) in our experiment be given as $\mathcal{P}^{t-1} = 4.05$ and $\mathcal{P}^t = 3.57$. The plaintext is encrypted as explained in Sect. 5.4. Similarly, for integrity checking since the size of C^t is 256 bits, the maximum size of N^0 will be $\log_2 256 = 8$ bits. The resultant C^t i.e. 256 + 8 bits is converted into their ASCII values generating 32 + 1 characters of ASCII which is the final cipher *Cipher*. Figure 3 shows the ciphertext generated for each of the plaintext from all 4 devices. The ciphertext *Cipher* is then sent to TC for decrypting.

Decryption TC, on receiving the ciphertext *Cipher*, now performs the reverse operation by first estimating the value $\mathcal{K}$ from the information provided by the device D_i. Value of Iteration *Itr* is calculated using similar approach with the help of the pre-shared session key $\mathcal{S}$. Ultimately ϑ is calculated through iteration on the chaotic map with the help of the control parameters, i.e. $\mathcal{K}$, $\mathcal{S}$ and *Itr*. TC extracts the plaintext by performing XOR operation of the previous known cipher C_{i-1}, the current cipher value C_i and the output produced by the chaotic map ϑ. The value obtained after the XOR operation is the plaintext $\mathcal{P}$ after converting to its decimal form, i.e. the value of 4.05 is successfully decrypted by the TC.

7 Security Analysis

Accomplishment of anonymity Anonymity ensures that even if the attacker A eavesdrops on any ongoing communication, he/she should not be able to detect the identity of either the sender or receiver of the intercepted message; i.e., the identity of a device D_i is completely anonymous. This is achieved in our protocol by the one-way property of hash functions which is the heart of our work in this paper. Intuitively, a one-way function is one which is easy to compute but difficult to invert. Thus, even if the hash proofs of the device D_i are intercepted by the attacker, it is impossible for him/her to extract the identity or the RFID of D_i, thus achieving device anonymity.

Accomplishment of the device authentication Our security protocol is based on the MHT where the values at each node are the hash of RFID tags. Since the RFID tags uniquely identifies a device, the generated hash values are also unique. In our protocol, any attacker attempting to initiate communication with the TC cannot forge the RFID of the authentic device and thus cannot deliver the accurate proofs. Moreover, even if the attacker in some way intercepts the RFID of the device, it is impossible to get hold of the hash values of all its siblings that constitute the valid proof. In this way, TC authenticates a legitimate device and prevents unwanted communication from untrusted third parties.

Accomplishment of data integrity To achieve data integrity, we have incorporated a mechanism where the ciphertext includes few bits for integrity checking. Suppose the ciphertext C be 11000110. The number of 0s is 4 or $N^{(0)} = 100$. Then, *Cipher* would be *Cipher* = 11000110|100 (where | signifies the division of *Cipher* into C^t and its 0 count $N^{(0)}$). Now if the ciphertext 11000110 were tampered by an attacker to 11000100 by changing the seventh bit to a 0, the value of $N^{(0)}$ being the same, the cipher would then be *Cipher* = 11000100|100. For the *Cipher* to be a valid codeword, the count $N^{(0)} = 100$ would also have to be changed to 101 because we now have 5 0s, not 4. But this requires changing a 0 to a 1, something that is forbidden. If the codeword were changed to 11000110|110 by altering $N^{(0)}$, then C would have to be changed so that it had 6 0s instead of 4. Again, this requires changing a 0–1 which is not possible. In this way, our algorithm guarantees data integrity.

Accomplishment of the data confidentiality Since our protocol does not require any exchange of keys, there is no way an untrusted third party can intercept the values and forge identity. Both the parties individually calculate the control parameters for the chaotic map, generate the pseudorandom number and estimate the final ciphertext. Therefore, the entire process of our proposed architecture is highly confidential and safe from tampering.

Resistance to replay attacks Having intercepted previous communication, the attacker can replay the same message of the sender to pass the verification process of the system. In our proposed approach, the value of $\mathcal{S}$ is completely random and is sent once in the beginning of the session. Since this is the only parameter value that needs to be shared (the value of $\mathcal{K}$ is known to both parties and does not require sharing and *Itr* can be calculated using $\mathcal{K}$ and $\mathcal{S}$ individually at both ends), an attacker cannot generate the initial conditions for the chaos map and thus cannot forge identity to intercept and resend the same message twice. However, if an attacker intercepts the most recent session key $\mathcal{S}$ and tries to send messages, TC will easily detect the obsolete value of $\mathcal{S}$ and declare it invalid. Since $\mathcal{S}$ is valid for one session only, any attempt to reuse its value is easily detected. In this way, our proposed authentication protocol can resist replay attacks.

Resistance to forgery attacks An attacker may also attempt to use the RFID of any legal validated device to pass the verification process of the TC. In that case, the attacker needs to construct a valid request message *REQ* with valid proofs to pass the TC's verification. However, to do that, he/she needs to not only the hash of the RFID but in addition the individual hash values of all the sibling nodes in its path to the TC, i.e. apart from $H(RFID)$, other proofs that include the ϕ values calculated for every node j at level i as $\phi_{i,j} = H(\phi_{i-1,2j-1}||\phi_{i-1,2j})$, which is quite impossible for him/her to figure out as these are the unknown secrets and therefore an attacker cannot convince TC of its identity. In this way, our proposed scheme can resist forgery attacks.

8 Conclusion

This paper proposes a security protocol by combining the advantage of both Merkle hash tree and chaotic cryptography. Specifically, we have modified the traditional MHT to suit an IoT environment by utilizing the RFID tags for constructing the tree. The chaotic cryptography is based on our proposed sinusoidal quadratic chaotic map whose dynamics and chaotic properties have been thoroughly tested. Our algorithms use lightweight computations that is well suited for the resource-constrained IoT devices. Experimental and security analysis proves the effectiveness of our algorithms and its resilience to security attacks.

Acknowledgements This work is partly supported by the Ministry of Electronics & Information Technology (MeitY), Government of India, under the Visvesvaraya PhD Scheme for Electronics & IT (PhD-PLA/4(71)/2015-16) and DST-Water Technology Initiative (WTI) (DST/TM/WTI/2k16/45(C)) dated 28 September 2016, Government of India.

References

1. Guo, B., Zhang, D., Yu, Z., Liang, Y., Wang, Z., Zhou, X.: From the internet of things to embedded intelligence. World Wide Web **16**(4), 399–420 (2013)
2. Lampropoulos, K., Denazis, S.: Identity management directions in future internet. IEEE Commun. Mag. **49**(12), 74–83 (2011)
3. Weber, R.H.: Internet of things-new security and privacy challenges. Comput. Law Secur. Rev. **26**(1), 23–30 (2010)
4. Liu, C., Ranjan, R., Yang, C., Zhang, X., Wang, L., Chen, J.: Mur-dpa: top-down levelled multi-replica merkle hash tree based secure public auditing for dynamic big data storage on cloud. IEEE Trans. Comput. **64**(9), 2609–2622 (2015)
5. Xu, K., Ma, X., Liu, C.: A hash tree based authentication scheme in sip applications. In: IEEE International Conference on Communications, ICC'08. IEEE (2008), pp. 1510–1514 (2008)
6. Niaz, M.S., Saake, G.: Merkle hash tree based techniques for data integrity of outsourced data.' In: GvD, pp. 66–71 (2015)

7. Wang, W., Si, M., Pang, Y., Ran, P., Wang, H., Jiang, X., Liu, Y., Wu, J., Wu, W., Chilamkurti, N., Jeon, G.: An encryption algorithm based on combined chaos in body area networks (2017). [Online]. Available: http://www.sciencedirect.com/science/article/pii/S0045790617324138

8. Wang, X.-Y., Zhang, Y.-Q., Bao, X.-M.: A colour image encryption scheme using permutation-substitution based on chaos. Entropy 17(6), 3877–3897 (2015)

9. Akhshani, A., Akhavan, A., Mobaraki, A., Lim, S.-C., Hassan, Z.: Pseudo random number generator based on quantum chaotic map. Commun. Nonlinear Sci. Numer. Simul. 19(1), 101–111 (2014)

10. Avaroğlu, E.: Pseudorandom number generator based on arnold cat map and statistical analysis. Turk. J. Electr. Eng. Comput. Sci. 25(1), 633–643 (2017)

11. Moreira, F.J.S.: Chaotic dynamics of quadratic maps. IMPA (1993)

12. Abundiz-Pérez, F., Cruz-Hernández, C., Murillo-Escobar, M., López-Gutiérrez, R., Arellano-Delgado, A.: A fingerprint image encryption scheme based on hyperchaotic rössler map. Math. Probl. Eng. 2016 (2016)

13. Murillo-Escobar, M., Cruz-Hernández, C., Abundiz-Pérez, F., López-Gutiérrez, R.M.: Implementation of an improved chaotic encryption algorithm for real-time embedded systems by using a 32-bit microcontroller. Microprocess. Microsyst. 45, 297–309 (2016)

14. Méndez-Ramírez, R., Arellano-Delgado, A., Cruz-Hernández, C., Abundiz-Pérez, F., Martínez-Clark, R.: Chaotic digital cryptosystem using serial peripheral interface protocol and its dspic implementation. Front. Inf. Technol. Electron. Eng. 19(2), 165–179 (2018)

15. Hassan, W.H., et al.: Current research on internet of things (iot) security: a survey. Comput. Netw. 148, 283–294 (2018)

16. Hou, J., Qu, L., Shi, W.: A survey on internet of things security from data perspectives. Comput. Netw. 148, 295–306 (2018)

17. Yao, X., Han, X., Du, X., Zhou, X.: A lightweight multicast authentication mechanism for small scale iot applications. IEEE Sens J 13(10), 3693–3701 (2013)

18. Ning, H., Liu, H., Yang, L.: Cyber-entity security in the internet of things. Computer p. 1 (2013)

19. Cai, X., Wang, Y., Zhang, X., Luo, L.: Design and implementation of a wifi sensor device management system. In: 2014 IEEE World Forum on Internet of Things (WF-IoT). IEEE, pp. 10–14 (2014)

20. Gope, P., Hwang, T., et al.: Untraceable sensor movement in distributed iot infrastructure. IEEE Sens. J. 15(9), 5340–5348 (2015)

21. Gope, P., Hwang, T.: Bsn-care: A secure iot-based modern healthcare system using body sensor network. IEEE Sens. J. 16(5), 1368–1376 (2016)

22. Loi, F., Sivanathan, A., Gharakheili, H.H., Radford, A., Sivaraman, V.: Systematically evaluating security and privacy for consumer iot devices. In: Proceedings of the 2017 Workshop on Internet of Things Security and Privacy. ACM, pp. 1–6 (2017)

23. Makhdoom, I., Abolhasan, M., Abbas, H., Ni, W.: Blockchain's adoption in iot: the challenges, and a way forward. J. Netw. Comput. Appl. 125, 251–279 (2019). [Online]. Available: http://www.sciencedirect.com/science/article/pii/S1084804518303473

24. Mookherji, S., Sankaranarayanan, S.: Traffic data classification for security in iot-based road signaling system. In: Soft Computing in Data Analytics, pp. 589–599. Springer, Berlin (2019)

25. Matheu-García, S.N., Hernández-Ramos, J.L., Skarmeta, A.F., Baldini, G.: Risk-based automated assessment and testing for the cybersecurity certification and labelling of iot devices. Comput. Stand. Interfaces 62, 64–83 (2019)

26. Merkle, R.C.: A certified digital signature. In: Conference on the Theory and Application of Cryptology, pp. 218–238. Springer (1989)

27. Li, H., Lu, R., Zhou, L., Yang, B., Shen, X.: An efficient merkle-tree-based authentication scheme for smart grid. IEEE Syst. J. 8(2), 655–663 (2014)

28. Lawande, Q., Ivan, B., Dhodapkar, S.: Chaos based cryptography: a new approach to secure communications. BARC newsletter, vol. 258, no. 258 (2005)

A Quantitative Methodology for Business Process-Based Data Privacy Risk Computation

Asmita Manna, Anirban Sengupta and Chandan Mazumdar

Abstract The imminent introduction of the Data Protection Act in India would make it necessary for almost all enterprises, dealing with personal data, to implement privacy-specific controls. These controls would serve to mitigate the risks that breach the privacy properties of user data. Hence, the first step toward implementing such controls is the execution of privacy risk assessment procedures that would help elicit the privacy risks to user data. All user data are processed/managed by one or more business processes. Hence, assessment of privacy risks to user data should consider the vulnerabilities within, and threats to, corresponding business process. It should also consider different perspectives, namely business, legal and contractual needs, and users' expectations, during the computation of data privacy values. This paper proposes such a comprehensive methodology for identifying data privacy risks and quantifying the same. The risk values are computed at different levels (privacy property level, business process level, etc.) to help both senior management and operational personnel, in assessing and mitigating privacy risks.

Keywords Business process · Data privacy · Personally identifiable information · Privacy risk assessment · Privacy threat · Privacy vulnerability

1 Introduction

The imminent introduction of the Data Protection Act in India [1] would make it necessary for almost all large-scale as well as mid-scale enterprises, dealing with

A. Manna (✉)
Department of Computer Science and Engineering, Jadavpur University, Kolkata, India
e-mail: asmita.nag@gmail.com

A. Sengupta · C. Mazumdar
Department of Computer Science and Engineering, Centre for Distributed Computing, Jadavpur University, Kolkata, India
e-mail: anirban.sg@gmail.com

C. Mazumdar
e-mail: chandan.mazumdar@gmail.com

© Springer Nature Singapore Pte Ltd. 2020
R. Chaki et al. (eds.), *Advanced Computing and Systems for Security*,
Advances in Intelligent Systems and Computing 996,
https://doi.org/10.1007/978-981-13-8969-6_2

personal data, to implement privacy-specific controls. These controls would serve to mitigate the risks that breach the privacy properties of user data, in general, and personally identifiable information (PII), in particular [2]. Hence, the first step toward implementing such controls is the execution of privacy risk assessment procedures that would help elicit the privacy risks to user data. Privacy risk assessment comprises of steps for identifying, analyzing and evaluating the potential harm that may be caused to user data if threats exploit vulnerabilities of the corresponding assets [2]. Here, assets refer to the entities that extract, process and store user data. They are of two types—primary assets and supporting assets [2, 3]. Primary assets of an enterprise are data and business processes, while, hardware, software, network, personnel, site and organizational structure are referred to as supporting assets. Evaluating risk of an enterprise involves evaluation of risks to each individual asset and application of a procedure to combine those risks to obtain a comprehensive risk value.

Historically, lots of research have been undertaken to assess *security risks* to assets [4]. Similarly, several researchers have channelized their efforts toward privacy impact assessment (PIA) [2]; but, most of these papers either focus on legal and organizational aspects or consider the risk assessment of hardware, software, network or information assets only. However, there is another class of risks that emanate from the existing vulnerabilities of business processes that handle data having privacy requirements. To the best of our knowledge and understanding, no work has so far tried to evaluate data privacy risks that occur due to vulnerabilities within, and threats to, business processes of an enterprise. Moreover, it is important to consider different perspectives, namely business, legal and contractual needs, and users' expectations, while deriving privacy risks. Aim of this research is to fill these gaps by proposing a comprehensive methodology for identifying the privacy risks owing to business processes and quantifying the same. The proposed methodology utilizes some of the notations and computation techniques used by Bhattacharjee et al. for deriving *security risk* values of business processes [5].

Rest of the paper is organized as follows. Section 2 presents a survey of related work. Section 3 briefly states the structure of a business process. Section 4 describes the proposed privacy risk assessment methodology and illustrates it with the help of a case study. Section 5 discusses the usefulness of the methodology. Finally, Sect. 6 concludes the paper.

2   Related Work

In this section, we discuss some of the published risk assessment techniques in the domain of user data privacy.

Mulle et al. [6] proposed a BPMN-based approach [7] for modeling privacy as a part of security in business processes and also proposed a methodology for implementing the same. Labda et al. [8] extended BPMN to propose a model of privacy-aware BPMN. Abu-Nimeh et al. [9] proposed an alteration in the existing security risk assessment techniques of SQUARE [10] to address privacy risks. The suggestion is to combine current risk assessment techniques with some PIA models [2].

Shapiro [11] proposed an integration anonymization framework within a privacy risk model. This framework is a data flow process where specific manipulations are undertaken either for making the data closer to be fed for anonymization or for actually anonymizing it. The author has developed a privacy risk model comprising of the proposals of Nissenbaum [12] and Solove [13]. The main drawback of this approach is that it only considers anonymization, while ignoring other privacy properties. Thus, it may suit the needs of some specific applications only.

In [14], the authors have proposed a method for enabling better choice for privacy settings based on a risk analysis framework. Users' privacy preferences and expectations are taken as input and, based on a specific risk value, perceived privacy risks are identified. The drawback of this approach is, though background knowledge has been considered, history of data disclosure, role of system architecture and context have not been taken into account.

Pellungrini et al. [15] have provided a methodology to evaluate privacy risk in retail data. They have defined the data formats for representing retail data, a privacy framework for calculating privacy risk and some possible privacy attacks for these types of retail data. The major drawback of this methodology is that it caters to retail data only and cannot be utilized for any other kind of private data.

Wagner and Boten [16] expressed privacy risk value as a distribution instead of an average score. They proposed a systematic methodology for the quantification of impact and likelihood of risk. These values are then combined to obtain a risk distribution for the enterprise as a whole. However, the authors have not provided any specific method for combining these values.

In [17], the authors have suggested a refinement approach to reuse the result of privacy risk analysis for integration with privacy by design. The proposed three-phase methodology allows comparison among different architectures with respect to privacy risks. In the first phase, generic harm trees are constructed using system components, risk sources and risk events. Phase 2 takes the specific architecture as input, and the generic harm trees are refined accordingly. During Phase 3, all context-specific components are considered and harm trees are further refined. Once, the final harm trees are constructed, risk is calculated as per the methodology proposed by Le Métayer and De [14].

Besides research papers, there are some standards that specify guidelines for privacy impact assessment and list the controls that may be implemented to protect privacy properties of user data. Notable among them are ISO/IEC 29134:2017 [2], ISO/IEC 29151:2017 [18] and NIST SP 800-53 rev. 4 [19].

It can be seen from the above discussion that existing papers are either domain-specific or consider privacy requirements from a single source, that is business needs only. There is a lack of methodologies that can combine the privacy requirements emanating from multiple sources, namely business, legal and contractual needs, along with customers' specific expectations. In this paper, we have attempted to fill this research gap by proposing a methodology that covers all three risk sources and also considers their relative importance while computing the privacy risk value.

3 Business Process

An enterprise achieves its business goals with the help of a set of well-defined business processes (a sample business process is shown in Fig. 1). All user data are processed/managed by one or more business processes. For example, data pertaining to a customer's credit history are processed by the loan management process of a bank if the customer applies for a loan. Similarly, a patient's medical records are processed by the analytical test processes of a diagnostic center during test report generation; the same medical records are also handled by the claims management process of a healthcare organization if the patient files a payment claim. Hence, assessment of privacy risks to user data should incorporate the potential harm that may emanate from corresponding business process. Moreover, it is apparent that the same user data may have different privacy risk values pertaining to different business processes which handle that data.

Computation of risk usually comprises of methods for assigning/deriving values of asset-specific properties, and combining them with relevant vulnerability and threat values, using qualitative or quantitative techniques [4]. In our approach, the privacy properties of a data item are first identified and evaluated. The vulnerabilities within, and threats to, corresponding business processes are then identified and their severities and impact values are calculated. Finally, these values are combined to derive the privacy risk value of the data item. These procedures are detailed in Sect. 4.

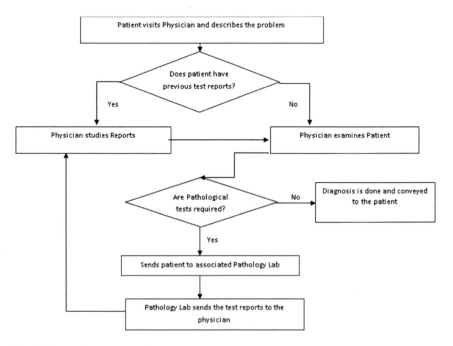

Fig. 1 Diagnosis management process

4 Privacy Risk Assessment Methodology

As stated above, our proposed methodology derives privacy risks by combining three parameters: privacy properties of data, vulnerabilities of business processes and threats to business processes. While doing so, our methodology considers the different perspectives of privacy that emanate from enterprise business needs, privacy laws and contractual obligations, and expectations of individual users/customers. The following subsections describe the steps of the risk assessment methodology.

4.1 Valuation of Data

Privacy of a data item hinges around the preservation of a set of properties, namely confidentiality, integrity, availability, anonymity, unlinkability, undetectability and unobservability [20]. Among these, confidentiality (C), integrity (I) and availability (A) are, primarily, security properties. However, research shows that in order to protect privacy of PII, preservation of C, I, A of the data item is of utmost importance [2]. Anonymity (AN) of a data item means one cannot identify its owner/principal within a given set [20]. Unlinkability (UL) is the inability to establish a link between two or more pieces of data items. Undetectability (UD) of a data item means that unauthorized subjects cannot determine whether it exists or not. Unobservability (UO) of a data item implies it cannot be *detected* by unauthorized subjects, and it is also *anonymous* for authorized subjects. In other words, unobservability is, in essence, a combination of anonymity and undetectability properties [20].

The proposed methodology evaluates a data item by computing the expected values of each of its properties. Values of the properties can be generated from enterprise business requirements, legal and contractual obligations, and expectations of users/customers. In the sample business process of Fig. 1, a patient's pathological test reports may have high C-value, A-value and AN-value from the patient's (customer) perspective. While from the enterprise's business angle, those reports may have high I-value. Again, existing laws may mandate high AN-value for the same reports.

Thus, it is obvious that the same data item may have different values of its privacy properties, from different perspectives. A comprehensive privacy risk assessment methodology should be able to address this distinction and compute risk values accordingly. Our technique is to identify the privacy needs of a data item based on business, legal and users' perspectives, for each privacy property.

Confidentiality Requirement

Confidentiality of data signifies that unauthorized users will not be able to access the data [20]. We propose C-value assignment on a 3-point scale, with "1" signifying negligible confidentiality requirement and "3" denoting high requirement. The need for assigning a particular C-value to a data item would be different from the viewpoints of business, legal authority and user/customer. Table 1 suggests a C-value

Table 1 Assigning values to privacy properties of data items

Value	Business requirement	Legal requirement	User requirement
3	Breach of (i) *confidentiality* may lead to irrecoverable financial loss and/or loss of customer base (ii) *integrity* may lead to impairment of business processes and/or erroneous outcomes that cause irrecoverable damage (iii) *availability* of data may lead to non-execution of business processes that cause irrecoverable damage (iv) *anonymity* may lead to irrecoverable loss of customer base and/or permanent blacklisting (v) *unlinkability* may lead to revelation of workflows and business secrets, thus causing irrecoverable financial loss and/or business closure (vi) *undetectability* may lead to irrecoverable financial loss, loss of trade secrets, and/or loss of customer base	Breach of (i) *confidentiality* may lead to litigation and/or breach of contract/SLA (ii) *integrity* may lead to litigation and/or breach of contract/SLA (iii) *availability* of data may lead to serious non-fulfillment of compliance obligations, leading to litigation and/or breach of contract/SLA (iv) *anonymity* may lead to litigation, cancellation of trade license and/or breach of contract/SLA (v) *unlinkability* may lead to non-compliance and hence serious litigation (vi) *undetectability* may lead to threat to national security, litigation and/or breach of contract/SLA	Breach of (i) *confidentiality* may lead to loss of proprietary/secret data, identity theft, or other irrecoverable loss to user (ii) *integrity* may lead to identity impairment, huge financial loss or other irrecoverable loss to user (iii) *availability* may prevent access to, and use of, PII, causing irrecoverable loss to user (iv) *anonymity* may lead to loss of reputation, embarrassment, identity theft, risk to life or other irrecoverable loss to user (v) *unlinkability* may lead to revelation of identity and/or sensitive information, causing serious loss to user (vi) *undetectability* may lead to loss of revelation of secrets, theft of sensitive data or other irrecoverable loss to user
2	Breach of (i) *confidentiality* may lead to significant financial loss and/or loss of customer base (ii) *integrity* may lead to erroneous/inconsistent outcomes that cause significant damage (iii) *availability* of data may lead to partial execution of business processes that cause significant damage (iv) *anonymity* may lead to significant loss of customer base and/or temporary blacklisting (v) *unlinkability* may lead to revelation of some workflows, thus causing significant financial loss and/or loss of customer base (vi) *undetectability* may lead to significant financial loss and/or loss of customer base	Breach of (i) *confidentiality* may lead to warning from legal authorities and/or third parties (ii) *integrity* may lead to warning from legal authorities and/or third parties (iii) *availability* of data may lead to significant non-fulfillment of compliance obligations (iv) *anonymity* may lead to warning from legal authorities and/or suspension of trade license (v) *unlinkability* may lead to partial non-compliance and some litigation (vi) *undetectability* may lead to potential warning from legal authorities and/or third parties	Breach of (i) *confidentiality* may lead to significant loss to user (ii) *integrity* may cause significant loss to user, including the pain of correcting/updating her data (iii) *availability* may cause partial inaccessibility to PII, causing significant loss to user (iv) *anonymity* may lead to significant loss to user (v) *unlinkability* may lead to significant loss to user (vi) *undetectability* may lead to significant loss to user
1	Breach may not lead to any significant loss	None	Breach may not lead to any significant loss

assignment scheme along with specific reasons for doing so. It may be noted that this is a representative set only and lists the most common concerns that may lead to specific value assignments. However, there may be other needs for assigning particular C-values and an enterprise can enhance/modify the contents of Table 1 suitably.

Integrity Requirement

Integrity of data refers to accuracy and completeness of data [20]. Like confidentiality, integrity may also assume values on a 3-point scale. Table 1 shows a sample I-value assignment scheme. This considers the fact that loss of integrity means loss of accuracy and/or completeness of data. For example, if the accuracy of a banking customer's PII is impaired, the bank will not be able to establish her identity, resulting in financial loss/mental trauma to the customer. As stated above, an enterprise can customize Table 1 based on its specific requirements.

Availability Requirement

Availability of data refers to accessibility and usability of data upon demand by authorized users [20]. Table 1 contains a sample A-value assignment scheme using a 3-point scale. For example, if a user is denied access to her PII, she may not be able to update the same on time, leading to blockage of essential services. It may be noted that availability and confidentiality requirements may have reciprocal relationship. For example, access to Aadhaar numbers of customers may be required for verifying their identities and addresses; however, this may lead to breaches of confidentiality wherein customers' contact numbers, biometric data, etc. are revealed to unauthorized entities. Hence, it is important to consider this aspect while assigning C-value and A-value to a data item, so that both of these requirements can be optimally addressed.

Anonymity Requirement

Unlike other privacy properties, anonymity is not about data but about the data subject [20]. If the anonymity of an individual is to be protected, the data set related to that individual should be protected in such a way that the data set alone cannot identify the individual. An individual can be identified from different sets of PII data; hence, an individual can have multiple anonymity values with respect to different sets of data. Table 1 provides some guidance on the assignment of anonymity values. It may be noted that confidentiality addresses protection of the contents of data, while anonymity strives to hide the link between data and its owner/principal. These should be considered during the assignment of values to these properties.

Unlinkability Requirement

Unlinkability is the inability to establish a link between two or more items of interest [20]. It may be so, that from a certain data set A, processed by business process P1, the data subject cannot be identified; similarly, it is also not possible to identify the subject from another data set B, processed by business process P2. However, it is possible to establish the identity of the subject from the combined data set of A and B. Hence, in order to protect the identity of the subject with respect to both the data sets, it should be ensured that A and B are *unlinkable*. Table 1 lists some specific requirements which may determine the assignment of unlinkability values.

Undetectability Requirement

Undetectability of a data item means that unauthorized subjects cannot determine whether it exists or not [20]. As undetectability does not mention any relationship between data and data owner, the onus of identifying this requirement lies primarily with the enterprise. It should be viewed as a part of the enterprise's overall strategy of protecting its information systems from attackers. It may be seen that while confidentiality aims to prevent unauthorized subjects from accessing a data item, undetectability attempts to hide the very existence of the data item. This means that the requirements that determine UD-values are similar to those that determine C-values of data items; this is shown in Table 1.

Unobservability Requirement

Unobservability of a data item means it is both *undetectable* for unauthorized subjects and *anonymous* for authorized subjects [20]. Hence, the assignment of UO-values can be done by combining the requirements of UN-values and AN-values.

Once the privacy values of a data item have been determined, vulnerabilities within the relevant business processes and the corresponding threats need to be identified. These are detailed in the following subsection.

4.2   Privacy-Specific Vulnerabilities Within Business Processes

By definition, vulnerability means inherent weakness in an entity that can be exploited by threats to breach data privacy [2, 3]. In this research, we focus on those vulnerabilities within business processes which, if exploited, can lead to a breach of privacy of user data. Such vulnerabilities arise due to improper configuration of processes or due to lack of checking and controls. Some examples of vulnerabilities within business processes are as follows:

- Owing to improper configuration, PII may be accessed by unauthorized processes;
- Incorrect processing of user data owing to lack of trained manpower;
- Absence of proper input validation may cause improper execution of business processes or generation of inaccurate outcome.

An enterprise should identify the vulnerabilities within its business processes and assess their seriousness. This is usually done by defining two properties: *severity* and *exploitability* [5]. Severity (Sev) of vulnerability indicates how badly the privacy of data may be impacted if threats exploit this vulnerability. Here, impact refers to the loss of privacy properties of data (confidentiality, integrity, anonymity, etc.). Breach of these properties may occur as follows:

Table 2 Assigning severity values to vulnerabilities within business processes

Sev	Interpretation
3	Vulnerability causes breach of privacy properties of PII
2	Vulnerability causes breach of privacy properties of non-PII items
1	Vulnerability does not cause any significant breach of data privacy properties

Table 3 Assigning exploitability values to vulnerabilities within business processes

Exp	Interpretation
3	Exploitation of vulnerability requires minimum or zero resources (tools, manpower, knowledge, time, etc.) and no authentication
2	Exploitation of vulnerability requires some resources and single level of authentication
1	Exploitation of vulnerability requires several resources and multiple levels of authentication

- Confidentiality is breached if unauthorized subjects can "study" the data item;
- Integrity is breached if unauthorized subjects can "alter" or "destroy" the data item;
- Availability is breached if unauthorized subjects can "alter" the data item or make it "inaccessible";
- Anonymity is breached if entities can "identify" the data owner/principal;
- Unlinkability is breached if it is possible to "identify" the relationship between multiple items of interest;
- Undetectability of data is breached if entities can "spot" the data item;
- Unobservability is breached if authorized subjects can "identify" the data owner/principal, and unauthorized subjects can "spot" the data item.

It is important to note that exploitation of a single vulnerability may result in a breach of one or more privacy properties in the manner listed above. For the purpose of this research, severity is computed on a 3-point scale, as shown in Table 2.

Exploitability (Exp) of vulnerability measures the ease with which the vulnerability can be exploited by threat(s) [5]. Table 3 suggests a scheme for assigning values to exploitability on a 3-point scale.

4.3 Privacy-Specific Threats to Business Processes

Threats are defined as perpetrators of unwanted incidents; they exploit vulnerabilities to cause a breach of data privacy [2, 3]. Traditionally, threats are categorized based on the source of initiation; for example, natural threats, environmental threats and threats induced by human beings [3]. After identifying vulnerabilities within business processes, an enterprise should analyze the corresponding threats that may exploit those vulnerabilities. All identified threats are not expected to cause harmful

Table 4 Assigning LOC values to threats

LOC	Interpretation
3	There have been past occurrences of the threat; the threat agent has all/most of the necessary resources, and authentication and authorizations required for exploiting the vulnerability
2	There have been past occurrences of the threat; the threat agent has some of the resources, and partial authentication and authorizations required for exploiting the vulnerability
1	There have been no past occurrences of the threat; the threat agent neither has sufficient resources, nor any authentication and authorization required for exploiting the vulnerability

incidents with the same probability. Hence, it is important to compute the *likelihood of occurrence* (LOC) of identified threats to comprehend the actual risk perception. It may be recalled that the *impact* caused by the exploitation of vulnerabilities by corresponding threats has already been considered during the computation of *severity* values. So, here we focus on the computation of LOC only.

In our proposed methodology, LOC value of a threat is computed by considering three factors: (i) past occurrences of incidents related to the threat; (ii) availability of resources (with the threat agent) required to exploit the corresponding vulnerability; and (iii) availability of authentication and suitable authorizations required to exploit the corresponding vulnerability. While the first factor is an estimate based on historical data, the remaining two factors consider the requirements of exploitability of vulnerability as listed in Table 3. We use a 3-point scale to determine LOC values; this is shown in Table 4. It may be noted that whether LOC will be assigned a value of 2 or 3 has to be determined based on two criteria: (i) the amount of resources and authentication/authorization available with the threat agent and (ii) the ease with which the threat agent can acquire the remaining resources and authentication/authorization. These would vary depending on the nature of the business process, threat agent, required resources and authentication/authorization.

4.4   Computation of Privacy Risk

As stated earlier, data privacy risk is a function of three parameters: (i) privacy value of data; (ii) vulnerability; and (iii) threat. Thus, after evaluating the privacy parameters of a data item, the vulnerabilities of relevant business processes and corresponding threats, risks to the data item can be assessed. Our methodology performs this assessment broadly in two steps. In the first step, threats are combined with corresponding vulnerabilities to obtain *privacy concern* value, while in the second step, privacy concern is combined with the values of privacy properties to determine the *privacy risk* value of data.

Privacy concern (PC) is computed as follows:

$$PC(t, v) = \text{ceil}((\text{LOC}(t) + \text{Exp}(v)) * \text{Sev}(v)/6) \qquad (1)$$

Here, $PC(t, v)$ denotes the concern that threat "t" will exploit vulnerability "v" within a business process; $\text{LOC}(t)$ denotes the likelihood of occurrence of threat "t"; $\text{Exp}(v)$ and $\text{Sev}(v)$ denote the exploitability and severity values of vulnerability "v", respectively. It can be seen that the sum $(\text{LOC}(t) + \text{Exp}(v))$ represents the potential of threat "t" being able to exploit the vulnerability "v", while the product of this sum with $\text{Sev}(v)$ denotes the probable impact that can occur if the incident actually occurs. The denominator 6 and *ceiling* function have been used to scale down the value in a manner that $\{2,…,6\}$ are mapped to "1", $\{8,…,12\}$ are mapped to "2" and $\{15, 18\}$ are mapped to "3". This ensures a uniform distribution of privacy concern values. Thus, $PC(t, v) \varepsilon \{1, 2, 3\}$.

The privacy concern values are computed for all threat–vulnerability (t-v) pairs of a business process. They are then grouped into six categories, corresponding to the privacy properties detailed in Sect. 4.1—confidentiality concern (C-concern), integrity concern (I-concern), availability concern (A-concern), anonymity concern (AN-concern), unlinkability concern (UL-concern) and undetectability concern (UD-concern). C-concern refers to all those privacy concern values that have been derived from such t-v pairs which can breach the confidentiality of a data item. Similarly, I-concern, A-concern, AN-concern, UL-concern and UD-concern denote values corresponding to integrity, availability, anonymity, unlinkability and undetectability properties of a data item, respectively. If a t-v pair can breach multiple privacy properties, then the pair will contribute privacy concern to multiple categories. For example, the threat "retrieval of discarded documents" can exploit the vulnerability "lack of secure disposal procedures" to breach confidentiality, unlinkability and undetectability properties of a data item. This gives rise to C-concern, UL-concern and UD-concern for the data item. It may be noted that "unobservability" has not been considered separately as it is a combination of anonymity and undetectability properties, as explained in Sect. 4.1.

The risk values of data items are obtained by combining the values of its privacy properties and privacy concerns of relevant business processes. This may lead to different types of risk values for the same data item, d_i: (i) d_i may have separate risk values pertaining to different privacy properties (confidentiality, anonymity, etc.); (ii) d_i may have separate risk values for the same privacy property but different sources of privacy requirement (business, legal or user requirement, as described in Sect. 4.1); and (iii) d_i may have separate risk values for the different business processes that process/use this data item. Hence, data privacy risk values for particular combinations of "requirement-property-business process" can be computed as:

$$^{xr}k\text{-risk}_{BP_j}(d_i) = \text{ceil}(^{xr}k\text{-value}(d_i)) * \max\left(k\text{-concern}(BP_j)/3\right) \qquad (2)$$

Here, xr denotes business requirement, legal requirement or user requirement; k-risk denotes confidentiality risk, anonymity risk, etc.; BP_j denotes the business process; and d_i denotes the data item. Hence, the term $^{xr}k\text{-risk}_{BP_j}(d_i)$ represents the value of k-risk to data item d_i for being handled by business process BP_j, from the perspective of x (business, laws and regulations or user). Similarly, ^{xr}k-value (d_i) denotes the k-value (C-value, AN-value, etc.) of d_i from the perspective of x-requirement, while, k-concern (BP$_j$) implies the value of k-concern (C-concern, AN-concern, etc.) of business process BP$_j$. For a business process BP$_j$, there can be multiple t-v pairs that can breach privacy property k of data item d_i, leading to multiple k-concern (BP$_j$) values. We consider the maximum among these values and multiply it with the k-value of d_i to obtain the risk value. The denominator 3 and *ceiling* function have been used to scale down the value uniformly, that is {1, ..., 3} are mapped to "1", {4, 6} are mapped to "2" and {9} is mapped to "3". Thus, ^{xr}k-risk$_{BP_j}(d_i)$ ε {1, 2, 3}.

Using the above notation, the *anonymity risk* to data item d_i, which is handled by business process BP$_j$, from the perspective of *users* will be computed as follows:

$$^{ur}\text{AN-risk}_{BP_j}(d_i) = \text{ceil}\left(^{ur}\text{AN-value}(d_i) * \max\left(\text{AN-concern}(BP_j)\right)/3\right) \quad (3)$$

Other types of risks can be computed in a similar manner. The risk values can be aggregated to obtain summarized values for better management and reporting. The following measure provides data privacy risk value by consolidating different sources of requirement:

$$k\text{-risk}_{BP_j}(d_i)$$
$$= \text{ceil}\left(\alpha * {}^{br}k\text{-risk}_{BP_j}(d_i) + \beta * {}^{lr}k\text{-risk}_{BP_j}(d_i) + \gamma * {}^{ur}k\text{-risk}_{BP_j}(d_i)\right) \quad (4)$$

where $0 \le \alpha, \beta, \gamma \le 1$ and $(\alpha + \beta + \gamma) = 1$.

Here, k-risk$_{BP_j}(d_i)$ denotes the value of k-risk to data item d_i for being handled by business process BP$_j$, from the combined perspective of business, legal and user requirements. The weights α, β and γ can be adjusted by an enterprise according to the priorities it associates with each of the sources of data privacy requirement. This may depend on several factors like stringency of laws and regulations, religious sentiments of users, etc. For example, if an enterprise assigns values 0.2, 0.4 and 0.4 to weights α, β and γ, respectively, the integrity risk to data item d_i pertaining to business process BP$_j$ will be given by:

$$I\text{-risk}_{BP_j}(d_i)$$
$$= \text{ceil}\left(0.2 * {}^{br}I\text{-risk}_{BP_j}(d_i) + 0.4 * {}^{lr}I\text{-risk}_{BP_j}(d_i) + 0.4 * {}^{ur}I\text{-risk}_{BP_j}(d_i)\right) \quad (5)$$

It may be seen from Eqs. (2) and (4) that k-risk$_{BPj}(d_i)$ ε {1, 2, 3}. Another measure can be proposed that computes privacy risk values by combining the risks that occur due to *different* business processes of an enterprise that handle a particular data item. This is computed as follows:

$$^{xr}k\text{-risk}(d_i) = \text{ceil}\left(\Sigma_j\left(\omega_y * {}^{xr}k\text{-risk}_{BP_j}(d_i)\right)\right), \text{ where } 0 \le \omega_y \le 1 \text{ and } \Sigma\omega_y = 1 \quad (6)$$

Here, $^{xr}k\text{-risk}(d_i)$ represents the value of k-risk to data item d_i from the perspective of x (business, laws and regulations, or user), considering *all* applicable business processes of the enterprise. The weights ω_y can be adjusted by the enterprise based on the importance of associated business processes. For example, if a business process is critical for the operations of the enterprise, higher weight may be assigned to it. This is owing to the fact that risks emanating from that business process need to be dealt with on an urgent basis to avoid jeopardizing business operations. It can be seen from Eqs. (2) and (6) that $^{xr}k\text{-risk}(d_i) \, \varepsilon \, \{1, 2, 3\}$.

The risk pertaining to a particular privacy parameter of data, considering *all* sources of requirements and associated business processes, can be derived from either Eq. (4) or Eq. (6). This is shown below:

$$k\text{-risk}(d_i) = \text{ceil}\left(\Sigma_j\left(\omega_y * k\text{-risk}_{\text{BP}_j}(d_i)\right)\right), \text{ where } 0 \leq \omega_y \leq 1 \text{ and } \Sigma \, \omega_y = 1 \tag{7a}$$

$$k\text{-risk}(d_i) = \text{ceil}\left(\alpha * {}^{br}k\text{-risk}(d_i) + \beta * {}^{lr}k\text{-risk}(d_i) + \gamma * {}^{ur}k\text{-risk}(d_i)\right), \tag{7b}$$

where, $0 \leq \alpha, \beta, \gamma \leq 1$ and $(\alpha + \beta + \gamma) = 1$

Equations (7a) and (7b) have been derived from Eqs. (4) and (6), respectively. It is obvious that if the weights of Eq. (7a) match with those of Eq. (6), and the weights of Eq. (7b) match with those of Eq. (4), then $k\text{-risk}(d_i)$ will have same values for both the computation methods. Also, $k\text{-risk}(d_i) \, \varepsilon \, \{1, 2, 3\}$.

Finally, the consolidated privacy risk of a data item, considering *all* parameters, can be computed as follows:

$$\text{risk}(d_i) = \text{ceil}\left(\Sigma_k\left(\delta_y * k\text{-risk}(d_i)\right)\right), \text{ where } 0 \leq \delta_y \leq 1 \text{ and } \Sigma \delta_y = 1 \tag{8}$$

Here, the weights δ_y can be adjusted based on the relative importance of the privacy properties. For example, in case of PII, confidentiality, integrity and availability may be accorded maximum weights; undetectability may be assigned lesser weight, while anonymity and unlinkability may be assigned least weights. It can be seen that $\text{risk}(d_i) \, \varepsilon \, \{1, 2, 3\}$.

Thus, this is how our proposed methodology assesses privacy risk of user data.

4.5 Case Study

Let us consider a case where the risk to a "patient's pathological test reports" is computed using the above methodology. Table 5 shows a sample value assignment for the various privacy properties. It may be noted that C-value and A-value have been assigned such that the requirements are not in conflict with each other. Similarly, *confidentiality* has been given higher priority than *undetectability* in the case study.

Now, let us consider two threat–vulnerability pairs that can affect the data item:

Table 5 Values of privacy properties of patient's pathological test reports

	BR	LR	UR
C-value	2	3	2
I-value	3	3	3
A-value	3	1	1
AN-value	3	3	3
UL-value	1	1	1
UD-value	1	1	1

i. The threat "data erasure/modification due to operators' mistakes" can exploit the vulnerability "absence of modification/deletion restriction on data which should not be modified," resulting in a breach of the integrity of the data item;

ii. The threat "billing department can view the laboratory reports along with patients' details" can exploit the vulnerability "no view restriction of patient's data" resulting in a breach of confidentiality, anonymity and undetectability of the data item.

For the first pair, both exploitability and severity of the vulnerability are high (values are 3 each). Similarly, the likelihood of occurrence of the threat can be considered as medium (value is 2). Hence, by Eq. (1), $PC(t1,v1) = \text{ceil}((2+3) *3/6)$ = 3. Similarly, $PC(t2,v2) = ((2+2) * 3/6) = 2$.

Risks can be calculated using Eq. (2) as follows:

$$^{br}I\text{-risk}_{BP_1}(d_i) = \text{ceil}((3^*3)/3) = 3; \quad ^{lr}I\text{-risk}_{BP_1}(d_i) = \text{ceil}((3^*3)/3) = 3;$$
$$^{ur}I\text{-risk}_{BP_1}(d_i) = \text{ceil}((3^*3)/3) = 3$$
$$^{br}C\text{-risk}_{BP_1}(d_i) = \text{ceil}((2^*2)/3) = 2; \quad ^{lr}C\text{-risk}_{BP_1}(d_i) = \text{ceil}((3^*2)/3) = 2;$$
$$^{ur}C\text{-risk}_{BP_1}(d_i) = \text{ceil}((2^*2)/3) = 2$$
$$^{br}AN\text{-risk}_{BP_1}(d_i) = \text{ceil}((3^*2)/3) = 2; \quad ^{lr}AN\text{-risk}_{BP_1}(d_i) = \text{ceil}((3^*2)/3) = 2;$$
$$^{ur}AN\text{-risk}_{BP_1}(d_i) = \text{ceil}((3^*2)/3) = 2;$$
$$^{br}UD\text{-risk}_{BP_1}(d_i) = \text{ceil}((1^*2)/3) = 1; \quad ^{lr}UD\text{-risk}_{BP_1}(d_i) = \text{ceil}((1^*2)/3) = 1;$$
$$^{ur}UD\text{-risk}_{BP_1}(d_i) = \text{ceil}((1^*2)/3) = 1$$

This is how the risks are calculated using our proposed methodology.

5 Discussion

The methodology described in this paper identifies and computes risks that can potentially cause harm to user data privacy when they are accessed/processed by an enterprise. The process begins by assigning values to different privacy properties of data like confidentiality, integrity, availability, anonymity, unlinkability and undetectability. Table 1 describes the suggested value assignment schemes of these properties. Then, the vulnerabilities within business processes are identified and their severity and exploitability values are computed. Also, the threats that can exploit these vulnerabilities, to breach the identified privacy properties, are analyzed and their *LOC* values are calculated. Finally, all of these values are combined to derive risk values corresponding to different privacy properties of data from the points of view of business, legal and user requirements. The methodology also describes techniques for deriving summary risk values by aggregating the individual scores. While summary values would aid senior management of an enterprise to take informed decisions on privacy investment, granular scores would help technical personnel implement privacy controls at the correct places.

The proposed methodology will help an enterprise to select proper risk mitigation techniques. These can be prioritized based on the importance of particular data items, criticality of privacy properties, and requirements of business, law enforcement agencies and users. It is possible to identify the business processes that are the biggest contributors to privacy risk; such processes can be controlled on a priority basis. Besides, the threat–vulnerability pairs, which can potentially breach a particular critical privacy property, can be easily identified by following the proposed methodology. This would help in implementing targeted controls that can mitigate the impact of particular (potentially dangerous) *t-v* pairs effectively.

In this paper, all values have been defined on 3-point scales. This has been done to maintain uniformity and present a simple, implementable approach for enterprises. Discussions with system managers and implementers have led us to believe that overtly complicated valuation schemes are time-consuming, costly and difficult to implement. Since assessment of privacy risks is a continuous process that has to be executed at regular intervals, adoption of a simple technique would be more practicable for enterprises. The qualitative interpretation of a 3-point risk scale would be as follows: "3" means "high risk" and needs immediate mitigation; "2" means "medium risk" and may need low-cost controls for remediation; "1" means "low risk" and may not need any controls immediately.

However, our methodology provides enough flexibility such that the values of different parameters may be assigned on higher granularities, if an enterprise so desires. For example, LOC values may be assigned on a 5-point scale as follows: "5" means several incidents have occurred in the past, and *all* resources and authentication/authorizations are available with the threat agent; "4"—several past incidents, availability of critical resources and most of the authentication/authorizations; "3"—some past incidents, availability of some critical resources and some of the authentication/authorizations; "2"—very few past incidents, non-availability of critical resources and availability of a few authentication/authorizations; "1"—no past incidents, non-availability of resources and authentication/authorizations. It is important to note that in case the point scales are changed, the various equations would need to be modified in order to scale down the risk values suitably. Another important aspect is the identification of vulnerabilities within, and relevant threats to, business processes. This generally varies between enterprises and enterprise sectors and should be diligently carried out by the privacy risk management team.

6   Conclusion and Future Work

A methodology for assessing privacy risks of user data has been presented in this paper. The paper begins by discussing some of the published works on privacy risk and modeling, along with their pros and cons. It has been shown that existing techniques fail to address privacy requirements from the point of view of different stakeholders. We have detailed how such requirements can be captured and integrated into a comprehensive privacy risk assessment framework. Besides, the methodology provides means of computing risk values at different levels (privacy property level, business process level, etc.) that will help both senior management and operational personnel, in assessing and mitigating privacy risks.

We intend to enhance the proposed methodology by incorporating the needs of privacy principles (purpose binding, consent and choice, collection limitation, etc.) within the suggested valuation scheme [18]. We would also like to delve deep into the structure of a business process and fine-tune the risk computation by considering the contributions of component tasks [5]. We will develop algorithms and a tool based on the proposed methodology. Results of privacy risk assessment using the tool will help in validating the technique that will lead to further improvement of the mechanism. Another interesting extension could be the formulation of techniques for combining privacy risks with information security risks to derive the overall value of risk to user data. This could be useful in the preparation of reports and dashboards for viewing by senior management and making critical investment decisions.

References

1. The Personal Data Protection Bill: http://meity.gov.in/writereaddata/files/Personal_Data_Protection_Bill,2018.pdf (2018). Last accessed 22 Sept 2018
2. ISO/IEC 29134:2017: Information technology—Security techniques—Guidelines for privacy impact assessment, 1st edn. ISO/IEC, Switzerland (2017)
3. ISO/IEC 27005:2011: Information technology—Security techniques—Information security risk management, 2nd edn. ISO/IEC, Switzerland (2011)
4. Bhattacharjee, J., Sengupta, A., Barik, M.S., Mazumdar, C.: A study of qualitative and quantitative approaches for information security risk management. In: Gupta, M., Sharman, R., Walp, J. (eds.) Information Technology Risk Management and Compliance in Modern Organizations, pp. 1–20. IGI-Global, USA (2017)
5. Bhattacharjee, J., Sengupta, A., Mazumdar, C.: A Quantitative methodology for security risk assessment of enterprise business processes. In: Proceedings of the 2nd International Conference on Information Systems Security and Privacy (ICISSP), pp. 388–399. SCITEPRESS, Italy (2016)
6. Mulle, J., von Stackelberg, S., Bohm, K.: Modelling and transforming security constraints in privacy-aware business processes. In: Proceedings of the IEEE International Conference on Service-Oriented Computing and Applications, pp. 1–4. IEEE (2011)
7. Business Process Model and Notation: http://www.bpmn.org/. Last accessed 21 Sept 2018
8. Labda, W., Mehandjiev, N., Sampaio, P.: Modeling of privacy-aware business processes in BPMN to protect personal data. In: Proceedings of the 29th Annual ACM Symposium on Applied Computing. ACM, Republic of Korea (2014)

9. Abu-Nimeh, S., Mead, N.: Combining privacy and security risk assessment in security quality requirements engineering. In: AAAI Spring Symposium: Intelligent Information Privacy Management (2010)
10. Risk   Assessment   Guide—SQUARE: https://www.square.org.au/risk-assessment/risk-assessment-guide/. Last accessed 22 Sept 2018
11. Shapiro, S.S.: Situating anonymization within a privacy risk model. In: 2012 IEEE International Systems Conference SysCon, pp. 1–6 (2012)
12. Nissenbaum, H.: Privacy in Context: Technology, Policy, and the Integrity of Social Life. Stanford Law Books, Palo Alto (2009)
13. Solove, D.: Understanding Privacy. Harvard University Press, Cambridge (2010)
14. Le Métayer, D., De, SJ.: Privacy risk analysis to enable informed privacy settings. In: [Research Report] RR-9125, Inria—Research Centre Grenoble—Rhône-Alpes, pp. 1–24 (2017)
15. Pellungrini, R., Pratesi, F., Pappalardo, L.: Assessing privacy risk in retail data. In: Guidotti, R., Monreale, A., Pedreschi, D., Abiteboul, S. (eds.) Personal Analytics and Privacy. An Individual and Collective Perspective. PAP 2017. LNCS, vol. 10708. Springer, Cham (2017)
16. Wagner, I., Boiten, E.: Privacy risk assessment: from art to science, by metrics. In: Garcia-Alfaro, J., Herrera-Joancomartí, J., Livraga, G., Rios R. (eds.) Data Privacy Management, Cryptocurrencies and Blockchain Technology, DPM 2018, CBT 2018. LNCS, vol. 11025. Springer, Cham (2018)
17. De, S.J., Le Métayer, D.: A refinement approach for the reuse of privacy risk analysis results. In: Annual Privacy Forum, vol. 10518, pp. 52–830. Vienne, Austria (2017)
18. ISO/IEC 29151:2017: Information technology—Security techniques—Code of practice for personally identifiable information protection, 1st edn. ISO/IEC, Switzerland (2017)
19. NIST SP 800-53: Security and privacy controls for federal information systems and organizations, 4th edn. NIST, USA (2013)
20. Pfitzmann, A., Hansen, M.: A terminology for talking about privacy by data minimization: anonymity, unlinkability, undetectability, unobservability, pseudonymity, and identity management, v0.34. http://dud.inf.tu-dresden.de/Anon_Terminology.shtml. Last accessed 16 Sept 2018

Architectural Design-Based Compliance Verification for IoT-Enabled Secure Advanced Metering Infrastructure in Smart Grid

Manali Chakraborty, Shalini Chakraborty and Nabendu Chaki

Abstract Smart Grid is built on an already existing and functioning power grid system, while adding various components to collect data as well as monitor, analyze and control the grid. In order to ensure the proper execution of the grid functionalities, it is important to verify the compliance issues before integrating a component in the system. In this paper, we propose a two tier compliance verification model, to verify the architectural compliance for each component as well as the whole system. Besides, this model monitors the data flow between several components in the system to record new compliance rules and take actions accordingly. Thus, this model can provide a certain level of security in the system while maintaining the architectural compliance. In order to justify this model, we have used Event-B to design an example and showed that if we use this model to select primitive components, then the architectural design will comply with the main system, as well as, certain security problems are also avoided using this framework.

Keywords IoT · Smart grid · Compliance · CBSD · AMI

1 Introduction

Smart Grid is an initiative to completely re-structure the electrical power grid to meet the current and future requirements of its customers [1]. It is an emerging technology with various improvements over the existing power industry. It introduces several

M. Chakraborty · S. Chakraborty (✉) · N. Chaki
Department of Computer Science and Engineering, University of Calcutta, Kolkata, India
e-mail: chakrabortyshalini.96@gmail.com

M. Chakraborty
e-mail: manali.chakraborty@unive.it

N. Chaki
e-mail: nabendu@ieee.org

M. Chakraborty
Università Ca' Foscari, Via Torino, 153, 30172 Venezia, VE, Italy

© Springer Nature Singapore Pte Ltd. 2020
R. Chaki et al. (eds.), *Advanced Computing and Systems for Security*,
Advances in Intelligent Systems and Computing 996,
https://doi.org/10.1007/978-981-13-8969-6_3

new functionalities in addition to the existing power generation, transmission and distribution systems to provide an automated, reliable, efficient, distributed and cost-effective power delivery system [2]. To achieve those new functionalities, Smart Grid uses a communication network and multiple service-oriented applications.

1.1 Motivation

Since, Smart Grid is built on an already existing and functioning system, it is necessary to verify the compliance issues before integrating a component in the system. Architectural compliance is one of the most important concerns for Smart Grid. To ensure architectural compliance, the system designer has to guarantee the proper transformation of architectural rules in the design and implementation of a system. The architecture of a system should provide information regarding the design issues of the system, i.e., the component structures, relationship between components, functional and non-functional requirements to comply with the policies and regulations of the organization, etc. [1, 3]. In Smart Grid, developers import the components from outside. This makes them vulnerable for violating the architectural rules of a system.

1.2 State of the Art

Compliance often refers to the validation of a system against some legal policies, internal policies or some basic design facts [4]. There exist several good works in compliance checking for business process models. Compliance checking can be of two types [5]:

- *Forward compliance checking* This method works proactively and can be further classified as design-time compliance checking and run-time compliance checking. In compliance by design method, the system is developed, by taking into account the business rules. Thus, the system is designed in such a way that it can comply with the rules [6]. On the other hand, run-time compliance checking methods detect and terminate non-compliant processes before execution.
- *Backward compliance checking* This method is a reactive one. It monitors the existing system thoroughly to detect whether it violates any rules or not. If it does not comply, then corrective measures are taken to make it compliant.

Authors in the paper [7] proposed a design-time compliance checking approach by measuring the deviation of any business process models with respect to a compliance pattern. In [8], authors used Formal Contract Language (FCL) to formalize the semantics of compliance rules, as well as the violations of business processes against these rules. Another compliance checking framework is presented in [9]. In this paper, authors transform the business process models in finite-state machines using an algebraic modeling language: pi-calculus and the strategies or rules are also formalized using LTL. Then, they used OPAL: open process analyzer, to verify the synchronization between process models and compliance rules.

A compliance checking at runtime method is proposed in [10]. It uses the log files of process models as a reference and provides a quantitative difference of new models with respect to those reference models. Whenever it detects non-compliance, it alerts the system with specific information about the violations. However, this work only considers control flow-based compliance issues.

There exist some good works in both forward and backward compliance checking; however, majority of them concentrate on business process models based on traditional software development techniques. In this paper, we are trying to verify the compliance of component-based software development (CBSD) processes. In CBSD, the system is not developed from scratch [11], and components are integrated in an already existing system based on functionalities. Thus, the design-time compliance method does not suit CBSD, as we will be using a lot of off-the-shelf processes in the system. The forward compliance checking methods will increase the time complexity of the verification process, as it has to check the whole system, every time when it will execute after adding a small component. Besides, finding the violations will be quite difficult too. Hence, we propose *Compliance Verification while Integration* for CBSD.

1.3   Contribution

In this paper, we have proposed a framework for two tier compliance verification model. The primary objective of this model is to identify only those of the shelves primary components that comply with the organization's basic architectural design. The candidate components are verified against the compliance rules using three criteria: Entailment, Consistency and Minimality [12] and the best component is selected for integration. The first tier works with individual component integration. The primary job of this tier is to ensure that a composite component is integrated using only those primitive components whose combined effects comply with the predefined semantic effects of the activity as well as the compliance rules. The second tier of compliance verification works with individual traces of the system, and it concerns with the compliance through the execution trace of an individual component, that is, the cumulative compliance checking for each component. This phase ensures that the integrated component is consistent with its previous components.

1.4   Problem Statement

Let us consider Smart Grid as a component-based system $S = (C, \Psi)$, consisting of a non-empty set C of *Composite Components* and an *Dependency* function Ψ that defines the dependency of composite components of S. $\Psi_{C_{i,j}}$ shows that composite components C_i and C_j are dependent on each other.

Every C_i can be further defined as a collection of non-empty set P of *Primitive Components*, F of *Functions* that describes the objective of C_i and C_R of *Rules* that defines the compliance rules associated with C_i.

Then, we can define our problem statement in two steps:

1. $\forall C_i \in C, C_{iR} \iff \cup_i P_{iR}$
   i.e., for any composite component, the collective compliance rules of all its primitive components should comply.
2. $\forall C_i \in C, C_{iR} \iff \cup_j \Psi_{C_{i,jR}}$
   i.e., for any composite component, the collective compliance rules of all its dependent composite components should comply.

In order to verify our proposed framework, we have used Rodin [13], which is a toolset for system-level modeling and analysis using Event-B modeling concept. In Event-B, set theory is used as modeling notation and refinement strategies are applied at each step to incrementally design the whole system. Besides, it also verifies the consistency at each level of abstraction using mathematical proofs. Since, Event-B allows the integration of any component in the system only when the requirement of that component confirms with the whole system and also refines the existing functionalities of the system along with the process, it emerges as one of the most suitable approaches to model a system like Smart Grid. Thus, in this paper, we have used this method to model AMI using our proposed framework.

The rest of the paper is organized as described here: Sect. 2 discusses the architectural design of our proposed system. The compliance management technique is described in details in Sect. 3. Section 5 verifies the effectiveness of our proposed framework and finally the paper concludes in Sect. 6.

2 Architectural Design of the System

Main objective of architecture is to define the structure and behavior of the system. Besides, the architectural design includes *Pre_Compliance_Rules* and *Post_Compliance_Rules*. The compliance rules are derived from the design documents of the system. For any sub-system S, its corresponding architectural design can be defined as

$$AD = (C, \Delta_i, \delta_o, \Gamma_{\text{Post}}, \Gamma_{\text{Pre}})$$

where

1. C defines the total number of components in the system. In the context of this framework, a component is a task-specific modular part of a system, which provides interfaces to other components for data communication. A component may be replaced by one or more off-the-shelf components while building the system, which again may be replaced by multiple primitive components if satisfies their integrated requirements.

Besides, a component can be defined by its properties as

$$C = (I_{\text{in}}, I_{\text{out}}, F)$$

where

- I_{in} = Number of components from which it takes inputs.
- I_{out} = Number of components to which it sends it outputs.
- F = set of specific functions or tasks. Each function is expressed as a set of atomic propositions, where an atomic proposition is defined as a declarative sentence that is either true or false and cannot be simplified further without modifying the logical equivalence of the proposition. Each atomic proposition is expressed in first-order logic (FOL).

2. Δ_i is the interface matrix. It defines the relation between components with respect to interfaces. Δ_i can be expressed as an (m * m) matrix, where m is the total number of components in the system.

$$\Delta_i[p][q] = 1, \quad \text{if } C_p \text{ and } C_q \text{ are connected through}$$
$$\text{an interface (either } I_{in} \text{ or } I_{out})$$
$$= 0 \quad \text{otherwise.}$$

3. δ_o can be defined as a function, mapping $(c * \Phi)$ to (s), where

- $c \in C$ and denotes all the components which have an interface to other existing sub-systems in the system.
- $s \in S$ and denotes all the existing sub-systems of the system which are connected to this component.
- Φ denotes the set of events, which triggers the data and control information interchange.

δ_o describes the degree of backward compatibility of a system. It denotes the dependency between newly installed components with pre-existing sub-systems within the system.

4. Γ_{Post} defines a set of *Pre Compliance Rules* for each component.
   Pre Compliance Rules are requirements and constraints that the system should achieve before the execution of that component. These rules are prerequisite for each component. A component cannot function accordingly and provide the desired goal if the system violates any of these rules.

5. Γ_{Pre} defines a set of *Post Compliance Rules* for each component.
   Post Compliance Rules for each composite component defines the requirements and constraints that the system should achieve after the execution of that component. The end results of an individual component must satisfy these rules to ensure stability, maintainability and proper synchronization of the system.

3   Compliance Management

The compliance management layer is responsible for checking the compliance of each individual component and as well as the whole system. It also records new compliance rules through the development process of the system.

3.1   Compliance Checking

In order to verify the compliance for each component as well as the whole system, we propose a two tier compliance verification model. The first tier works with individual component integration. The primary job of this tier is to ensure that a composite component is integrated using only those primitive components whose combined effects comply with the pre-defined semantic effects of the activity as well as the compliance rules.

The second tier of compliance verification works with individual traces of the system, and it concerns with the compliance through the execution trace of an individual component, that is, the cumulative compliance checking for each component. This phase ensures that the integrated component is consistent with its previous components.

Let, $T = t_1, t_2, \ldots, t_n$ is the set of traces in any component-based system, where n is the total number of traces and each t_i consists of activities in sequential order and represents a unique trace.

Then, we can define the functions as:

$$Pre\_Comply : T \times \mathbb{N} \rightarrow \wedge_{\Gamma_{Pre}}$$
$$Post\_Comply : T \times \mathbb{N} \rightarrow 2^{\Gamma_{Post}}$$

where Γ_{Post} and Γ_{Pre} are defined in the previous section.

The function *Pre_Comply* returns a conjunctive normal form of compliance rules that an activity in a trace has to fulfil to comply with the rules and regulations of the organization.

On the other hand, the function *Post_Comply* gives the effects of an activity that should confirm the compliance rules.

Consider the system of Fig. 1,

$T = t_1, t_2, t_3, t_4$, where
$t_1 = A, B, C, H, I$
$t_2 = A, B, D, F, H, G, I$
$t_3 = A, B, D, G, F, H, I$
$t_4 = A, B, E, I$

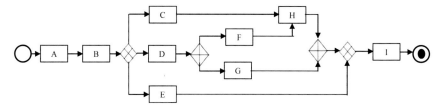

Fig. 1 Flow of activity of a system

Now,

$$\text{Comply}(t_1, 4) = C(x) \wedge C(y) \wedge C(z) \text{ and}$$
$$\text{Effect}(t_1, 4) = p, q, r$$

describe that the fourth activity in trace t_1, i.e., H has to comply with compliance rules: $C(x), C(y)$ and $C(z)$ and as a result of their association the output of that activity should match with the effects: $\{p, q, r\}$.

Our main objective is to build a system using component-based software development strategy that ensures compliance checking while integrating components. For that, we assume each activity in the system as a composite component and every composite component is developed using one or more primitive components such that the new developed component can fulfil the desired objectives as well as comply with all the normative requirements of the organization.

First Phase—Compliance for Component Integration: Let us assume that we want to develop composite component B. Now, we already know that, B has to comply with *Compliance Rules*: $C(x) \wedge C(y) \wedge C(z)$ and the effects should match with the compliance effects: $\{p, q, r, s\}$. [As Comply $(t1, 2) = C(x), C(y), C(z)$ and Effect $(t1, 2) = p, q, r, s$].

We may have multiple primitive components available off-the-shelf that satisfy some subset of these conditions as well as some additional constraints. For example, let P1, P2, P3, P4 be four such off-the-shelf components that satisfy the effects $(p, q, \neg t)$, (r, s, t), (p, r) and (q, s), respectively. In order to build the entire system, we have to integrate only those primitive components which satisfy two conditions: Entailment and Consistency [12].

Entailment: The combined *Pre Compliance Rules* and *Post Compliance Rules* of primitive components should satisfy the *Pre Compliance Rules* and *Post Compliance Rules* of the entire composite component.

For example, primitive components $P1$ and $P2$ can be combined to satisfy all the four conditions (p, q, r, s). Similarly, $P3$ and $P4$ can also be combined to satisfy the system conditions. Thus,

$$(p, q, \neg t) \wedge (r, s, t) \vdash (p, q, r, s, \neg t, t).$$
$$(p, r) \wedge (q, s) \vdash (p, q, r, s).$$

Suppose composite component C is developed by integrating primitive components $P1, P2, \ldots, Pn$, then,

$$(P1.\textit{Post Compliance Rules} \wedge P2.\textit{Post Compliance Rules} \wedge \cdots \wedge$$
$$Pn.\textit{Post Compliance Rules}) \vdash C.\textit{ Post Compliance Rules}$$

and

$$(P1.\textit{Pre Compliance Rules} \wedge P2.\textit{Pre Compliance Rules} \wedge \cdots \wedge$$
$$Pn.\textit{Pre Compliance Rules}) \vdash C.\textit{Pre Compliance Rules}$$

Consistency: The combined effect of the primitive components should be logically consistent, i.e., they should not derive falsity.

For example, $P1$ and $P2$ satisfy all the system constraints but are mutually inconsistent as

$$(p, q, \neg t) \wedge (r, s, t) \models F$$

On the other hand, $P3$ and $P4$ are mutually consistent and entail the system conditions as well.

$$(p, r) \wedge (q, s) \models T$$

Thus, we can generalize this as

$$(P1.\textit{Post Compliance Rules} \wedge P2.\textit{Post Compliance Rules} \wedge \cdots \wedge$$
$$Pn.\textit{Post Compliance Rules}) \nvDash F$$

and

$$(P1.\textit{Pre Compliance Rules} \wedge P2.\textit{Pre Compliance Rules} \wedge \cdots \wedge$$
$$Pn.\textit{Pre Compliance Rules}) \nvDash F$$

Minimality: The minimality criterion tries to extract that particular set of primitive components that minimize a certain criterion or soft goal. Soft goals usually represent QoS parameters. To evaluate the minimal combination of primitive components, we need to annotate each primitive component with respect to that particular soft goal.

Once this is done, we find out all combinations of primitive components that satisfy entailment and consistency, and select that a particular combination which minimizes the selected soft goal. Then the compliance manager reports to the component management layer, and the component management layer integrates those primitive components.

Second Phase—Compliance Checking with Dependant Components: To ensure compliance through an entire execution trace in the system, we have to ensure that the post-execution effects of each component is consistent with the previous components in that trace, i.e., the immediate effects of component H should be consistent with the cumulative effects of H for every traces with H in it. As an example, H is a part of three traces in the diagram: t_1, t_2 and t_3.

Besides, it may sometime be possible to invoke some extra compliance rules and effects while integrating primitive components. In such cases, it is necessary to validate these new rules with respect to the cumulative rules and effects of each trace in which the components participate.

Thus, the immediate effect of H should be consistent with the cumulative effects of (A, B, C, H), (A, B, D, F, H) and (A, B, D, G, F, H).

Now, the cumulative rule and effect of each trace can be calculated using the union of individual rules and effects of each components in that trace. Thus, we can verify the compliance of the whole system by ensuring that the immediate effects and compliance rules of the collective primitive components are consistent and entailed with the cumulative rules and effects of the trace.

Algorithms: Algorithm 1 describes the process of primitive component selection for a particular composite component C. Suppose the algorithm takes N primitive components as input and considering entailment, consistency and minimality criteria, gives the best set of primitive components to achieve the functionality of C.

Algorithm 1: *Compliance_Checking*$(C, P_1, P_2, \ldots, P_n, N)$

Input: C: Structure of the composite component C.
      $P_1, P_2, \ldots, P_n$: Structure of N primitive components.
      N: number of primitive components to be considered.
Output: *Primitive_Component*: Best set of primitive components for composite component
      C.

1 initialization;
2 *Entailed_Components = Entailment_Checking*(C, P1, P2... Pn, N);
3 *Consistent_Components = Consistency_Checking (Entailed_Components)*;
4 *Primitive_Component = Minimality_Checking (Consistent_Components,*
  $M_1, M_2, \ldots, M_t, t)$;
  ```
/* Record new compliance rules                              */
/* "-" denotes the set difference operation.                */
```
5 *Additional_Pre_Compliance$_R$ules =*
  (*C.Pre_Compliance_Rules − Primitive_Component.Pre_Compliance_Rules*);
6 *Additional_Post_Compliance$_R$ules =*
  (*C.Post_Compliance_Rules − Primitive_Component.Post_Compliance_Rules*);

Algorithm 2 finds all possible subsets over Pis (for all $i = n$) which satisfies the entailment criteria and gives *Entailed_Components* set, and Algorithm 3 provides the lists from *Entailed_Components* set which satisfy the consistency criteria. Finally Algorithm 4 identifies the best possible set of primitive components that satisfies all of the three criteria, i.e., entailment, consistency and minimality.

Algorithm 2: $EntailmentChecking(C, P_1, P_2, \ldots, P_n, N)$

Input: C: Structure of the composite component C.
  $P_1, P_2, \ldots, P_n$: Structure of N primitive components.
  N: number of primitive components to be considered.
Output: *Entailed_Component*: Structure of sets, contains the union of two or more
  primitive components which holds the entailment criteria.

1 initialization;
2 Find all possible subsets from $(P_1, P_2, \ldots, P_n)$, excluding the null set
  $0 and all the singleton sets, i.e., P_1, P_2, \ldots, P_n$;
3 $t = 2^N - N - 1$;
4 **for** $i = 1$ *to* t **do**
5     $(Pre\_Compliance\_Rules) =$
  $\forall(primitive components in ith subset), \cup_i P_i.Pre\_Compliance\_Rules)$;
6     $(Post\_Compliance\_Rules) =$
  $(\forall(primitive components in ith subset), \cup_i P_i.Post\_Compliance\_Rules)$;
7 **end**
8 **if** $C.Pre\_Compliance\_Rules \in (P_i.Pre\_Compliance\_Rules))$ &&
  $(C.Post_Compliance_Rules \in (Subset_Pi.post_Compliance_Rules)$ **then**
9     $Structure Entailed_Components_k := Subset_Pi$;
10   k=k+1;
11 **end**

3.2 Record New Rules

While integrating the primitive components, it is sometimes necessary that the components, both primitive and composite, should comply with some new rules for successful execution. Thus, another function of compliance manager is to record new compliance rules at runtime and keep the business rule database up to date with each change in the system.

Algorithm 3: $Consistency_Checking(Entailed_Components)$

Input: *Entailed_Component*: Structure of sets, contains the union of two or more primitive
  components which holds the entailment criteria.
Output: *Consistent_Component*: A structure of sets contains the union of two or more
  primitive components which holds both the entailment criteria and consistent
  criteria.

1 initialization;
2 count = number of elements in $Entailed_Components$;
3 **for** $i = 1$ *to* count **do**
4   **if** $(\forall i, \wedge_i Entailed_Components_i.Pre_Compliance_Rules) \not\models F$ &&
  $(\forall i, \wedge_i Entailed_Components_i.Post_Compliance_Rules) \not\models F$ **then**
5        $Consistent_Components = Entailed_Components_i$;
6   **end**
7 **end**

Algorithm 4: *Minimality_Checking(Consistent_Components)*

Input: *Consistent_Component*: Structure of sets, contains the union of two or more
      primitive components which holds the entailment and consistent criterias.
$M1, M2, \ldots, MT$: QoS parameters of individual primitive components.
$\delta M1, \delta M2, \ldots, \delta MT$: System Tolerability for each QoS parameter. It defines the level up to
which the system can perform efficiently without compromising the Quality of Service.
Output: *Primitive_component*: The best possible combination of primitive components
      which satisfies Entailment, Consistency and Minimality.

1   initialization;
2   count = number of elements in *Consistent_components*;
3   temp = 0;
4   **for** $i = 1$ *to count* **do**
5   **for** $j = 1$ *to T* **do**
6          $\lambda j = \delta Mj - Mj$ $//\lambda j$ gives the difference between system's tolerance and
           component's value for a QoS parameter.
7   *Consistent_Components$_i$.Minimality_Score$+ = \lambda j$*;
8   **end**
9   // sort the components according to their Minimality_Score. **if**
      Consistent_Components$_i$.Minimality_Score $> temp$ **then**
10   *temp = Consistent_Components$_i$.Minimality_Score*;
11   *Primitive_Component = Consistent_Components$_i$*;
12   **end**
13 **end**

3.3 Monitor the Integrated System for Compliance

Compliance of each primitive and composite component does not always imply that the whole system is also compliant with the business rules. Therefore, after checking the primitive and composite components for compliance, the compliance manager monitors the whole integrated system for compliance. A baseline is approved only when the system is compliant with the business policies.

4 AMI and IoT

Smart Grid is built around the existing grid, while adding different components that collects data at different points at the system and uses those data to optimize the efficiency of the system. Hence, the underlying communication network is the most crucial part of the Smart Grid, along with advanced metering infrastructure or AMI, where smart meters collect the usage details of each user and send it to data aggregators (DAs). These DAs will again send the aggregated data to data collection units (DCUs). Then, DCUs will transmit the data to meter data management system or MDMS, where the data is analyzed and used in various applications. The primary objective of this system is to provide a both-way communication between user and utilities, and based on this communication system, AMI provides several service-oriented applications, such as, smart billing, advanced forecasting, dynamic pricing and demand response systems. Besides, it also supports distributed generation of electricity in the grid.

AMI deploys various types of devices for monitoring, analyzing and controlling the grid, at various points, such as, powerplants, transmission line, transmission towers, distribution center, customer premises, etc. One of the main concern is the connectivity, automation and tracking of such large number of devices. Now, this type of connectivity of enormous number of devices is already done in the real life using IoT technology. Hence, IoT can be considered as the most suitable technology to transform the electricity grid. Figure 2 depicts the integration and communication of IoT system with the AMI.

In this paper, we have considered the application of distributed generation system in the AMI. Now, let us explain this application briefly. Suppose, some customers produce electricity in their homes, using solar panel or small bio-electricity generation systems, etc. Now, the customers will use this electricity for their daily usage, and if they produce excess electricity, then they can sell it to the grid. Figure 3 describes the working of this application.

Since, Smart Grid is an extension of the existing electrical grid, it will have to add new components to properly implement such applications. As an example, the

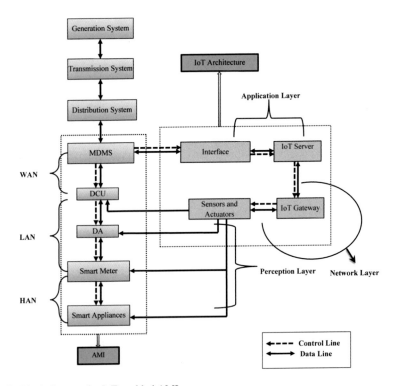

Fig. 2 Block diagram for IoT enabled AMI

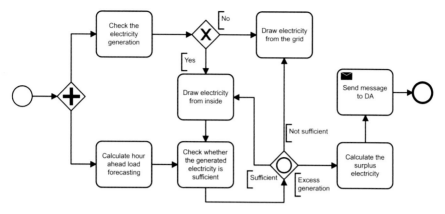

Fig. 3 Working principle for the distributed generation in AMI

smart meter has to add an additional component to predict hour-ahead demand of its owner based on some parameters, and the smart meter should also have a switch to toggle between two incoming electricity source: grid and home, etc.

Now, all of these components must comply with the architectural compliance of the existing system. Moreover, The integration of IoT makes this scenario a bit more complex.

5 Verification of the Proposed Framework

In order to verify our proposed framework, we have used Event-B modeling [2], which is a formal approach toward system-level modeling and analysis. Event-B works as a complete modeling approach for systems like Smart Grid, as at every step of design, along with modeling, it verifies the consistencies between levels and also checks the correctness of the functionalities. The advantage of Event-B is that it provides refinement at every stage of modeling. As a result, we can modify the system according to the new requirements which has come along with the selected primitive component.

The whole process is subdivided into three phases. In **Phase I**, we have identified the functional and non-functional requirements of the complete composite system. Our aim here is to check the correctness of these functionalities and we do that by modeling them using Event-B. In **Phase II**, we have disintegrated the composite component into several primitive components, each having individual functional and non-functional requirements. Then, using our algorithms we find the valid components. In our last phase, i.e., in **Phase III** we tried to model those collective primitive components with their new added requirements so that the new system can comply with the old system. We have used Rodin modeling tool [13] which is an Eclipse-based IDE for Event-B that provides support for refinement and proof for system modeling. We have considered the application described in Sect. 4 as an example.

5.1 Phase I

First of all, we have made a list of basic functional and non-functional requirements and checked their correctness with Event-B modeling.

Functional requirements:

- **FR_1**: The smart meter need to/must differentiate between the electricity coming from grid and that is coming from the solar panel.
- **FR_2**: The system must be able to monitor the electricity generation and thus check whether it is sufficient or not.
- **FR_3**: The system should be enabled to calculate the day-ahead demand of a household.
- **FR_4**: If the power coming from solar panel is not enough, the house can then buy electricity from the grid.
- **FR_5**: If there is excess power at the end of the day, the system should be able to sell.

Non-Functional requirements:

- **NFR_1**: Data integrity should be maintained.
- **NFR_2**: The system should maintain the relation between grid, generation unit and home owner.
- **NFR_3**: As the first source of electricity is solar panel, if for some reason the panel gets damaged, the system should be able to sense that and provide recovery measures by switching to the grid electricity immediately.

Now, Table 1 lists all the components that are required for the proper execution of this application.

Let us take the following assumptions:

- The average demand of household is y
- The capacity of the distributed electricity generation source is x
- Usually $x \geq y$

Table 1 Components of smart meter

Components	Responsibilities
Forecasting device (F)	Forecast the hour ahead demand of the household
Calculating device (C)	Calculate the capacity of the distributed electricity generation source
Monitor (M)	Controls the flow of electricity and the two different channels
Sensors (S1 and S2)	Enable the channels

When for some reason y increases to '$y + d$,' then following two scenarios can happen:

1. $x > (y + d)$: electricity from distributed electricity generation source is enough.
2. $x < (y + d)$: shortage of electricity.

Following are the actions taken for both of the scenarios.

- Step 1: If 'y' increases to '$y + d$,' M sends a request to the Grid.
- Step 2: Check the status between 'x' and 'y'.
- Step 3: If $x > (y + d)$, M kept the grid channel disabled.
- Step 4: If $x < (y + d)$, M disable the channel from distributed electricity generation source and enables the channel from the Grid.

In case of second scenario, the requirement F1 is needed and for that we first define a context with the set STATUS defining the two different statuses of the channels: enable and disable.

- Set: STATUS
- Constants: enable and disable
- Axioms & invariants:

  - $axm\_1 > Status=$ enable, disable
  - $inv\_1 > S1.enable$ if and only if S2.disable and vice versa.

Axioms define the context, whereas invariants are conditions which remain true despite the changes over in time. Now, we take three variables:

- *channel_sensor_sp*
- *channel_sensor_grid*
- *monitor_control*

In this connection, the monitor acts as an action, whereas the channel acts as a reaction. As we know, the reactions of the channels are strongly synchronized to the

Table 2 List of patterns

a	monitor_control
r.s	channel_sensor_sp
r.g	channel_sensor_grid
0	disabled
1	enabled
a_on	monitor_enable_channel
a_off	monitor_disable_channel
r.s_on	sp.channel_enabled
r.s_off	sp.channel_disabled
r.g_on	grid.channel_enabled
r.g_off	grid.channel_disabled

```
r.s_on
    when
        r.g = 0
        a = 1
    then
        r.s = 1
    end

sp.channel_enabled
    when
        channel_sensor_grid = disable
        monitor_control = enable
    then
        channel_sensor_sp = enable
    end
```
(a) solar panel channel enabled

```
r.g_on
    when
        r.s = 0
        a = 1
    then
        r.g = 1
    end

grid.channel_enabled
    when
        channel_sensor_sp = disable
        monitor_control = enable
    then
        channel_sensor_grid = enable
    end
```
(b) Grid channel enabled

```
r.s_off
    when
        a = 1
        r.g = 0
    then
        a = 0
        r.s = 0
        a = 1
        r.g = 1
    end

sp.channel_disable
    when
        monitor_control = enable
        channel_sensor_grid = disable
    then
        monitor_control = disable
        channel_sensor_sp = disable
        monitor_control = enable
        channel_sensor_grid = enable
    end
```
(c) solar panel channel disabled

```
r.g_off
    when
        a = 1
        r.s = 0
    then
        a = 0
        r.g = 0
        a = 1
        r.s = 1
    end

grid.channel_disable
    when
        monitor_control = enable
        channel_sensor_sp = disable
    then
        monitor_control = disable
        channel_sensor_grid = disable
        monitor_control = enable
        channel_sensor_sp = enable
    end
```
(d) Grid channel disabled

Fig. 4 List of Events

action of the monitor. Table 2 describes the patterns which are used to instantiate the situation:

This leads to the list of events as shown in Fig. 4, which are supposed to represent the reactions of the channels and from that we can show that the smart meter can differentiate between electricity coming from different sources.

5.2　Phase II

Suppose we have four primitive components to construct a composite component. Table 3 describes the functional and non-functional requirements of each primitive components. Now, we can observe that none of the primitive components is self sufficient to fulfil all the requirements. From the above four PCs, we can construct $2^4 = 16$ possible combinations of promising candidates: [(PM1), (PM2), (PM3), (PM4), (PM1, PM2), (PM1, PM3), (PM1, PM4), (PM2, PM3), (PM2, PM4), (PM3, PM4), (PM1, PM2, PM3), (PM1, PM2, PM4), (PM1, PM3, PM4), (PM2, PM3, PM4), (PM1, PM2, PM3, PM4)]. As per our algorithm, out of these subsets only (PM1, PM3) is the valid.

5.3　Phase III

The results of Phase II give us our valid set of components but it comes with two new requirements: F7 and NF4. Now, these two new requirements are included along with the valid set of component and thus we need to validate these two against our existing system. We can perform the validation process/we can prove the validation by modeling them with Event-B.

Let us take requirement NF4 which is an important requirement for PM3 that ensures the security of the excess electricity, i.e., the excess electricity cannot be sold outside without the consent of DA of that smart meter. This requirement limited the access of the excess electricity. Now in our basic requirements of the existing system, we do not have any security requirements like this. Hence, let us refine our integrated system with its new component and new requirement.

The first added component (excess electricity detector or EED) has two parts: sender and receiver. The sender part works as an alarm. When the smart meter has excess electricity that alarm is triggered and all smart meters within one-hop distance receive that signal. The DA under which that smart meter is registered also receives this signal. Thus, the information of excess electricity is known in the neighborhood.

Now as soon as DA receives this signal, it sends another signal to the sender smart meter with timestamp. In the meantime, other smart meters within one-hop distance also send signals to this smart meter for buying the electricity.

Table 3 Primitive components with functional and non-functional requirements

Primitive components	Functional requirement	Non-functional requirement
PM1	{F1, F2, F3}	{NF1, NF2}
PM2	{F4, F5, ^F2,F6}	{NF1, NF3, Nf4}
PM3	{F4, F5, F7}	{NF3, NF4}
PM4	{F2, ^F5, F3}	{NF3, ^NF2}

Now, the second added component is an array (A) where the first field contains the amount of total excess electricity and fields from that on contains the ids of all requesting smart meters. Within the time period, the smart meter holding excess electricity sends a response to the DA along with the array.

If the DA got the response within the defined timestamp, it determines the price of that electricity and also selects the smart meter to sell that electricity. Then, it sends another signal to the smart meter along with a packet containing the price and id of the buyer smart meter.

If the DA did not get a response from the sender smart meter within the time period defined by the timestamp, it assumes that smart meter is selling the electricity unethically and then can take necessary actions accordingly.

Now to model above scenario, we first define a context with the set STATUS defining the two different statuses of the channels: sending and receiving, and two different status for the timestamp: true and false.

- Set: STATUS
- Constants: sending, receiving, true and false
- Axioms & invariants:

  - axm_1: Status_channel= sending, receiving
  - axm_2: Status_timestamp= true, false

Now we take five variables:

- EED_signal
- DA_signal
- DA_timestamp
- Array_signal
- OHSM_signal (signal from one hop smart meters)

Invariants: inv_1: DA_signal=receiving if and only if DA_timestamp=true

Here the EED and DA are strongly synchronized but EED and other one-hop smart meters (OHSM) are weakly synchronized. We will use the following pattern as shown in Table 4, to instantiate the situation. Figure 5 gives the series of events for this situation.

At the end of these three phases, we have a verified composite system. Event-B uses refinement strategy and thus it is very easy and efficient to model new components with an already existing system. Here, at first we verify and model the composite system's requirements with Event-B and later on when we got our valid set of primitive components, we just need to refine the previous model in order to include new requirements, we do not need to remodel it from beginning. As an example in Phase 3, we successfully include NF4 into our existing system, and in the same way if we can include F7, then it is proved that the primitive components are totally compatible with the existing system. Also since all the requirements are well verified and modeled, we achieved our goal that is complete compliance verification.

Table 4 List of patterns for Phase 3

e	EED_signal
d	DA_signal
a	array_signal
ts	DA_timestamp
s	OHSM_signal
0	sending
1	receiving
t	true
f	false
ack	acknowledgement of excess electricity
s_rec	OHSM_signal_receives

```
ack
    when
        e = 0
    then
        s = 0
        d = 0
        ts = t
    end

ack
    when
        EED_signal = sending
    then
        OHSM_signal = sending
        DA_signal = sending
        DA_timestamp = true
    end
```

(a) Acknowledge excess electricity

```
s_rec
    when
        ts = t
        a = 0
        d = 1
    then
        ts = f
        d = 0
        a = 1
        s = 1
    end

s_rec
    when
        DA_timestamp = t
        Array_signal = sending
        DA_signal = receiving
    then
        DA_timestamp = false
        DA_signal = sending
        Array_signal = receiving
        OHSM_signal = receiving
    end
```

(b) Sell excess electricity

Fig. 5 List of Events in Phase 3

6   Conclusions

The proposed framework in this paper is responsible for securing the compliance rules of the system. The algorithm for compliance checking further uses three sub-algorithms to ensure three important aspects: entailment, consistency and minimality. The main concern in minimality checking is to identify the QoS parameters or soft goals which are relevant to the system and quantified each primitive component with respect to those soft goals. However, identify system-specific QoS variables and quantify them according to system specifications are altogether an independent topic for research, and we would like to explore this topic as a future extension of this work.

We have used Event-B to verify our framework. For future scope, we are trying to use Event-B method to model a better security system. In this paper, we have modeled one non-functional requirement which limited the access of excess electricity but we are trying to model an approach where we can stop unethical hacking of electricity.

Acknowledgements We would like to acknowledge Council of Scientific & Industrial Research (CSIR), Government of India, and the project "ADditive Manufacturing & Industry 4.0 as innovation Driver (ADMIN 4D)", for providing the support required for carrying out the research work.

References

1. IEEE Recommended Practice for Architectural Description of Software-Intensive Systems. IEEE Std 1471-2000, pp. i–23 (2000). https://doi.org/10.1109/IEEESTD.2000.91944
2. Abrial, J.R.: Modeling in Event-B: System and Software Engineering, vol. 1st. Cambridge University Press, New York, NY, USA (2010)
3. Herold, S.: Checking architectural compliance in component-based systems. In: ACM Symposium on Applied Computing (2010)
4. Lohmann, N.: Compliance by design for artifact-centric business processes. In: 9th International Conference on Business Process Management, pp. 99–115 (2011)
5. Kharbili, M.E., de Medeiros, A.K.A., Stein, S., van der Aalst, W.M.P.: Business process compliance checking: current state and future challenges. MobIS **141**, 107–113 (2008)
6. Sackmann, S., Kahmer, M., Gilliot, M., Lowis, L.: A classification model for automating compliance. In: 10th IEEE Conference on E-Commerce Technology and the Fifth IEEE Conference on Enterprise Computing, E-Commerce and E-Services, pp. 79–86 (2008). https://doi.org/10.1109/CECandEEE.2008.99
7. Sadiq, S., Governatori, G., Milosevic, Z.: Compliance checking between business processes and business contracts. In: IEEE International Enterprise Distributed Object Computing Conference (EDOC'06), pp. 221–232 (2006)
8. Ghose, A.K., Koliadis, G.: Auditing business process compliance. In: Proceedings of the International Conference on Service-Oriented Computing (ICSOC-2007). Volume 4749 of Lecture Notes in Computing Science, pp. 169–180 (2007)
9. Liu, Y., Muller, S., Xu, A.K.: Static compliance checking framework for business process models. IBM Syst. J. **46**(2), 335–361 (2007)
10. Rozinat, A., van der Aalst, W.M.P.: Conformance checking of processes based on monitoring real behavior. Inf. Syst. **33**(1), 64–95 (2008)
11. Crnkovic, I.: Component-based software engineering—new challenges in software development. J. Comput. Inf. Technol. **3**, 151–161 (2003)

12. Darimont, R., Lamsweerde, A.V.: Formal refinement patterns for goal-driven requirements elaboration. In: Proceedings 4th ACM Symposium on the Foundations of Software Engineering (FSE4), San Francisco, pp. 179–190, Oct 1996
13. Abrial, J.R., Butler, M., Hallerstede, S., Hoang, T.S., Mehta, F., Voisin, L.: Rodin: an open toolset for modelling and reasoning in Event-B. Int. J. Softw. Tools Technol. Transf. (STTT) Spec. Sect. VSTTE **12**(6), 447–466 (2010) (Springer, Berlin)

A Novel Approach to Human Recognition Based on Finger Geometry

Maciej Szymkowski and Khalid Saeed

Abstract Biometrics is one of the most important branches of science toward computer systems safety. One of the physiological traits is finger geometry as part of hand geometry. The purpose of this paper is to introduce an algorithm that can extract significant features from finger geometry by which human recognition can be achieved. All samples were obtained with the device created by the authors. The software part of the proposed approach consists of simple image processing methods that improve image quality before feature extraction, feature vector creation and classification. A total of 150 samples from fifty users were included in the analysis of the proposed system. Identity connected with each sample can be automatically provided by the system with an accuracy of 83%. The results were calculated on the basis of k-nearest neighbors algorithm and different distance calculation metrics: Euclidean, Manhattan and Chebyshev.

Keywords Finger geometry · Human recognition · Biometrics · Image analysis · Classification

1 Introduction

Nowadays, biometrics is one of the most important branches of science. It allows to recognize user only by his measurable traits not by what he remembers or has (e.g., card or document). Biometrics can be divided into two main groups: physiological biometrics that consists of traits acquired after birth, for example, retina, iris or fingerprint; and behavioral which is connected with methods based on what we can learn, for instance, keystroke dynamics or signature.

Finger geometry is also a physiological trait. It is not as popular as others (e.g., fingerprint). In the literature, one can easily find different approaches connected with

M. Szymkowski (✉) · K. Saeed
Faculty of Computer Science, Bialystok University of Technology, Bialystok, Poland
e-mail: m.szymkowski@pb.edu.pl

K. Saeed
e-mail: k.saeed@pb.edu.pl

© Springer Nature Singapore Pte Ltd. 2020
R. Chaki et al. (eds.), *Advanced Computing and Systems for Security*,
Advances in Intelligent Systems and Computing 996,
https://doi.org/10.1007/978-981-13-8969-6_4

57

hand geometry although there are only a few papers that deal with human recognition based on finger geometry. In most of them, recognition accuracy is not measured or is really low. Moreover, mostly finger geometry is used in multimodal biometrics systems. Due to the descriptions of analyzed systems [1, 2], it has to be claimed that finger geometry is only an ancillary feature in them. It is used only to confirm the result obtained with the main feature.

In this paper, we consider finger geometry for human recognition. Despite the fact that it can guarantee high accuracy level for human recognition, it is not used in the everyday-use systems. The most important advantages of this feature are ease of obtain and high uniqueness.

Unfortunately, like any biometrics trait, finger geometry has also its drawbacks. In this case, one of them is that different position of the finger during the trait retrieval may result in a completely different feature vector. It may lead to incorrect user recognition. In addition, this trait is not difficult to spoof. Despite these disadvantages, it still can provide high recognition accuracy. To ensure additional protection against fraud, modification on the hardware and software sides is needed. For instance, we can check whether blood flows in finger. It can be done with Eulerian Video Magnification [3].

In this work, we present a novel approach to human recognition based only on finger geometry. We are considering multiple parameters that can be obtained from human finger like finger widths and curvature angle. Various information obtained from the finger contour allows to reduce the possibility of human recognition with spoofed trait. Moreover, these data can provide much higher accuracy than the approach based only on one parameter.

This document is organized as follows: in the first section, the authors describe known approaches to process finger and hand geometry images. In the second, the proposed solution is presented. The third section contains information about performed experiments, especially about different parameters in the classification method. Finally, the conclusions and future work are given.

2 State of the Art

In the literature, we can find different approaches using biometrics for human recognition. The authors were looking for various algorithms and solutions where finger geometry was used. Most of the interesting works were connected with hand geometry only a few papers were about finger geometry acquisition and processing.

The authors of [4] proposed an algorithm to hand geometry processing. In this work, the fully automated multimodal approach based on hand geometry and vein system was presented. The method used for hand geometry processing was based on a few simple image processing algorithms like grayscale conversion, binarization and median filtering. After preprocessing, feature extraction was done—in the case of this system, specific information about each finger was obtained. The authors claimed that their approach has huge accuracy although they do not provide any information about the database on which this value was calculated. Moreover, the most important disadvantage of the proposed solution is too complicated feature vector that describes each sample.

Another interesting algorithm was presented in [5]. In this case, the authors combined hand geometry with finger geometry. It means they created two feature vectors for human description—one for hand geometry and one for the second measurable trait. In the paper, it was mentioned that one of the problems is that user can put a hand under the camera in different ways which will cause that feature vectors could be different. It can lead to wrong recognition. In the paper, decision algorithm based on Manhattan distance was also presented. The accuracy of the proposed approach was calculated. The main drawback of this method is too complicated image processing algorithm for hand geometry.

In [6], the authors used hand widths for human recognition. They proposed simple feature vector that describes each sample. This approach can be divided into three main parts: image preprocessing with simple algorithms (like binarization or median filtering), feature extraction with wrist detection and as the final one, classification. In the case of this approach, also k-nearest neighbors with different metrics were used to determine human identity. Despite the huge accuracy of the approach, information about the database is incomplete, and also there are no data about research duration.

In the literature, we can also find different approaches that use: discrete cosine transform [7], artificial neural networks [8, 9], convolutional neural networks [10] and machine learning [11]. It has to be claimed that all of these algorithms are connected with hand geometry. There are no suitable approaches for finger geometry.

We can conclude that the novelty of our work lies in the independent use of finger geometry for human recognition. In most of the known approaches, finger geometry is always used alongside hand geometry. This causes that the results are obtained with multimodal biometrics system where finger geometry is only an ancillary feature.

3 Proposed Algorithm

In this chapter, the authors present the complete algorithm that was worked out. The proposed approach is based on experiences obtained during previous works [12, 13]. Simultaneously, the authors are working on fully automated biometrics system based on three features obtained from human finger. These are fingerprint, finger veins and finger geometry. We will use finger geometry to check whether identity was properly evaluated on the basis of fingerprint and finger veins. Finger geometry processing and classification algorithm presented in this chapter will be a part of the recently mentioned biometrics system.

The algorithm consists of a few steps based on simple image processing algorithms. The activity diagram of the proposed approach is presented in Fig. 1. The first step of the approach is image acquisition.

The finger geometry image is obtained with the device created for this aim. It consists of three LEDs covered by the piece of white paper placed below the finger

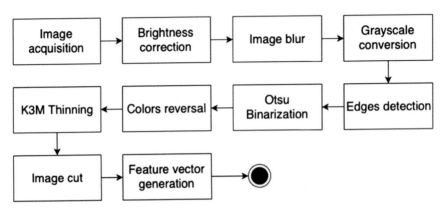

Fig. 1 Activity diagram of the proposed algorithm

and Tracer Prospecto Cam with additional lightning placed above the finger. The scheme of the system is presented in Fig. 2. The image acquired with the described device is presented in Fig. 3.

The first step in image processing was connected with image brightness correction. It was done due to the necessity of nonsignificant information removal that is

Fig. 2 Device scheme that was used to collect samples [14]

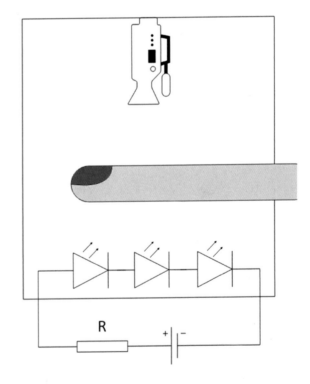

Fig. 3 Original image obtained with the proposed device

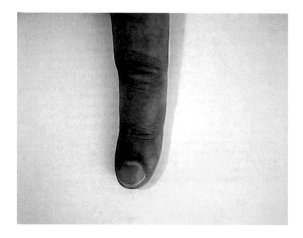

represented by lines in the center of the finger. The result of this operation is presented in Fig. 4.

The next step in image processing algorithm was connected with image blur. This procedure was used because we only need information about finger contour—other parts of the finger that are still visible in the image are not necessary in the case of human recognition. This operation was done in the grayscale image. Conversion to grayscale was done with green channel value. The image after conversion to grayscale and filtration with blur mask is presented in Fig. 5.

The fourth stage of the proposed approach was connected with edge detection. In this approach, Sobel algorithm with horizontal and vertical masks was used. Another approach that was tested during the experiments was Prewitt algorithm although it did not provide satisfactory effects. The result obtained after Sobel algorithm with horizontal and vertical masks is presented in Fig. 6.

Fig. 4 Image after brightness correction

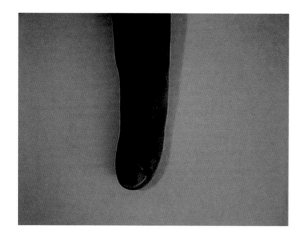

Fig. 5 Image after grayscale conversion and filtration with blur mask

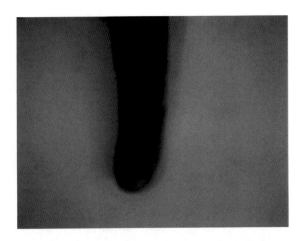

Fig. 6 Image with detected edges by Sobel algorithm

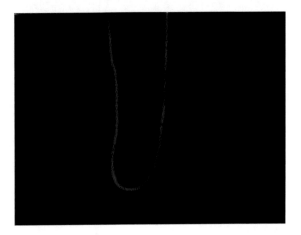

The fifth step in our approach is binarization. This procedure was used because we need more visible contour of the finger. After the previous step, it was not as clear as it is needed in the next steps of image processing. Binarization was done with Otsu algorithm [15]. The effect gained with this procedure is presented in Fig. 7a. The next stage was connected with color reversal. It means that the contour of the finger is marked with black color and the foreground is in white. This method was needed due to the requirements of thinning algorithm used in the following step. The result of color reversal is presented in Fig. 7b.

The next step of image processing was thinning. We took into consideration different approaches like KMM [16], K3M [17] and Zhang-Suen thinning algorithm [18]. The experiments showed that the most precise results were obtained with K3M method. It means that the finger contour is clearly visible and small number of additional points is presented in the image. The result of this procedure is presented in Fig. 8a. The last step of image processing part was connected with image cut. It

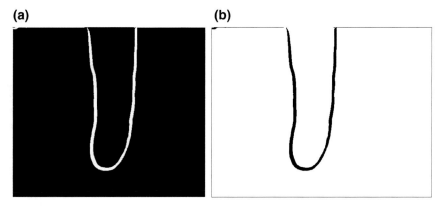

Fig. 7 Image after binarization with Otsu algorithm (**a**) and after color reversal (**b**)

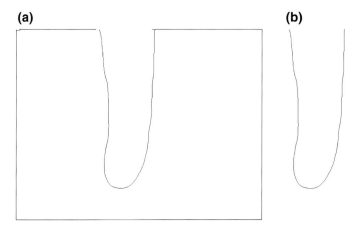

Fig. 8 Image after skeletonization with K3M algorithm (**a**) and after additional points removal (**b**)

was done because we needed to remove any additional points that are visible in the form of picture frame. The procedure used for this aim was 1-px frame removal. The image after this procedure is presented in Fig. 8b.

The final stage of our method is connected with feature extraction. As it was mentioned at the beginning of the algorithm description, two parameters were taken into consideration: vector consisting of finger widths and finger curvature angle. For the first of them, we are measuring the width in all possible levels. We start from the upper part of the finger and finish at the fingertip. The feature vector consists of values different from zero. The second value we use is finger curvature angle. In this case, linear interpolation was used to calculate this value. Calculation is done on the basis of two extreme points—one is the center of the finger base and the second is fingertip. The curvature angle is calculated as in (1).

$$\alpha = arctg\left(\frac{y_1 - y_0}{x_1 - x_0}\right) \tag{1}$$

where

α—is a finger curvature angle
x_0, x_1—horizontal coordinates of extreme points
y_0, y_1—vertical coordinates of extreme points.

In the next chapter, we present the results of the experiments connected with different weights of each of the parameters we extracted in the last step of the proposed method.

4 Experiments

The significant part of this work is connected with performed experiments. All of them were done on the database consists of 50 unique users. Each of them was described by 3 samples. The duration of the work was 2 months. To calculate identification accuracy, we used k-nearest neighbors algorithm and each of the parameters was given weight which was also taken to calculate the distance between two samples. The calculations were done as in (2).

$$d(A, B) = w1 \cdot d_w(A, B) + w2 \cdot d_k(A, B) \tag{2}$$

where

$d(A, B)$—it is a distance between feature vector A and feature vector B
$w1$—it is a weight with which distance between two width vectors is taken into final calculations
$d_w(A, B)$—it is a distance between two sets of widths (one set belongs to sample A and the second to B)
$w2$—it is a weight with which distance between two curvature angles is taken into final calculations
$d_k(A, B)$—it is a distance between two curvature angles (one belongs to sample A, and the second to sample B).

Experiments were done for different weights that satisfy the condition (3). Each of them was changed by 0.05. In Figs. 9 and 10, we present the results including weights that were selected for each parameter.

$$w1 + w2 = 1 \tag{3}$$

The best observed result is 83%. It was visible in huge amount of the cases with different weights. We can conclude that it can be observed when curvature angle has higher weight as well as when width vector was more important. These results

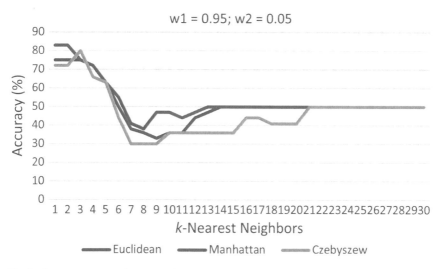

Fig. 9 System accuracy when $w1 = 0.95$ and $w2 = 0.05$

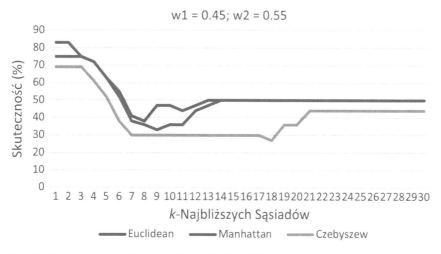

Fig. 10 System accuracy when $w1 = 0.45$ and $w2 = 0.55$

showed that selected measurable trait and classifier are commensurate with each other and allow human recognition on satisfactory level.

We also check the accuracies for different metrics: Euclidean, Manhattan and Chebyshev. The best results were obtained with Manhattan and Euclidean metrics. The worst results were observed when Chebyshev (otherwise known as the infinity metric) distance was used.

4.1 Comparison to Others' Solutions

In the literature, the authors did not find any other approaches connected with human recognition on the basis of finger geometry. However, there are a few solutions [2, 4–11] related to recognition based on hand geometry.

The authors would like to present undeniable advantages connected with usage of finger geometry rather than hand geometry. At the beginning, we would like to present a comparison between proposed approach and selected, the most interesting (in terms of design, efficiency and performance) algorithms accuracies. The summary of this examination is presented in Table 1.

The first advantage is easier use of acquisition device rather than in the case of hand geometry. We would like to highlight that it is much easier to put finger below the scanner than hand.

The second benefit is related to the finger geometry recognition efficiency. The authors observed that the accuracy is decreased in comparison with hand geometry although we have to claim that we do not have as much available parameters as from hand geometry. The conclusion that comes from this examination is that efficiency is not much decreased in comparison with hand geometry algorithms.

The third benefit is connected with time that is needed to obtain recognition results. The authors implemented three algorithms (all in Java programming language and C++) and prepared a summary that is presented in Table 2. Each experiment was run on the same computer (Intel Core i7, 16 GB RAM, 256 GB SSD and Mac OSX). Time was measured from the beginning of data processing till the decision about human identity was obtained.

We would like to highlight that the proposed approach allowed to get decision about human identity much faster rather than by the other three algorithms (based on different biometrics trait). Profit, in reference to the worst approach, was sevenfold. In comparison with the best one (from the other three algorithms), benefit in time

Table 1 Comparison between the proposed approach and others analyzed

Algorithm	Accuracy (%)
Bhatia and Guiral [5]—Hand geometry	97
Saxena et al. [6]—Hand geometry	97.4
El-Afy [9]—Hand geometry	98
Our proposed approach	83

Table 2 Comparison of time needed to obtain decision about human identity

Algorithm	Time
Bhatia and Guiral [5]—Hand geometry	45 s
Saxena et al. [6]—Hand geometry	1 min 10 s
El-Afy [9]—Hand geometry	43 s
Our proposed approach	10 s

equals 4.3. The result has shown that usage of simpler biometrics trait can provide huge earnings in the case of time needed for getting a decision.

In comparison with the other approaches [5, 6, 9], our idea is not only much faster but also more lightweight. By this conclusion, the authors would like to point out that in their approach only simple image processing algorithms were used to obtain the final result. We do not use any artificial intelligence methods due to the fact that it will be really hard to teach the neural network (or genetic algorithm) with satisfactory results on a small data set with not too much description parameters.

5   Conclusions and Future Work

In this work, an approach to human recognition based on finger geometry was presented. During experiments, two parameters from finger geometry were taken into consideration: vector of finger widths and finger curvature angle.

The results showed that it is possible to recognize the user with satisfactory accuracy level when more important is the vector of finger widths (it has higher weight) as well as when finger curvature angle influence is higher than finger geometry significance.

The described approach was implemented in real development environment. In our software, each step of the image processing can be displayed for the user. We can observe what is the current status of the image and how it looks like.

The proposed approach was tested on 50 unique users' database. Each of them was described by three samples. The highest accuracy obtained with the proposed approach was 83% for the considered rather large group of user in the database. This result shows that it is highly possible to properly determine human identity with his finger geometry for a specific small group of users. However, probably this trait when used as the only feature in a biometrics system would not be sufficient when the database is much larger. The differences between fingers from various users may be too small to properly evaluate human identity according to the features obtained from their finger geometry. The authors are working on the possibility of extracting more unique features in the finger and hope the near future will show more new results.

In this work, the authors presented an approach to classification with k-nearest neighbors algorithm. In the future, we would like to use soft computing method to deal with this task. We are considering convolutional neural networks and machine learning algorithms, for example, decision trees [19] or naïve Bayes classification [20].

As a future work, the authors would like to provide changes in the software and hardware to get much better results. Moreover, the authors work under the creation of fully automated multimodal biometrics system based on finger veins, fingerprint and finger geometry that will be implemented with Raspberry Pi. In this system, we will solve the problem of insufficient finger geometry accuracy in the process of human recognition.

Acknowledgements This work was supported by Grant S/WI/3/2018 from Białystok University of Technology and funded with resources for research by the Ministry of Science and Higher Education in Poland.

References

1. Le-qing, Z., San-yuan, Z.: Multimodal biometric identification system based on finger geometry, knuckle print and palm print. Pattern Recogn. Lett. **31**(12), 1641–1649 (2010)
2. Kang, B.J., Park, K.R.: Multimodal biometric method based on vein and geometry of a single finger. IET Comput. Vision **4**(3), 209–217 (2010)
3. http://people.csail.mit.edu/mrub/evm/. Accessed 11 Dec 2018
4. Tae Park, G., Kim, S.: Hand biometric recognition based on fused hand geometry and vascular patterns. Sensors **13**, 2895–2910 (2013)
5. Bhatia, A.K., Guiral, S.K.: Authentication using hand geometry and finger geometry biometric techniques. Int. J. Comput. Trends Technol. **4**(8), 2276–2781 (2013)
6. Saxena, N., Saxena, V., Dubey, N., Mishra, P.: Hand geometry: a new method for biometric recognition. Int. J. Soft Comput. Eng. **2**(6), 192–196 (2013)
7. Al-Ani, M.S., Rajab, M.A.: Biometrics hand geometry using discrete cosine transform (DCT). Sci. Technol. **3**(4), 112–117 (2013)
8. Faundez-Zanuy, M., Elizondo, D.A., Ferrer-Ballester, M.A., Travieso, C.M.: Authentication of individuals using hand geometry biometrics: a neural network approach. Neural Process. Lett. **26**, 201–216 (2007)
9. El-Afy, E.-S.M.: Automatic identification based on hand geometry and probabilistic neural networks. In: 2012 IEEE 5th International Conference on New Technologies, Mobility and Security, Istanbul, Turkey, Proceedings, 7–10 May 2012
10. Afifi, M.: 11 K Hands: gender recognition and biometric identification using a large dataset of hand images. arXiv: 1711.04322v9 [cs.CV], 17 Sep 2018
11. Jiang, X., Xu, W.: Contactless Hand Recognition (2006)
12. Szymkowski, M., Saeed, K.: A multimodal face and fingerprint recognition biometrics system. In: Saeed, K., Homenda, W., Chaki, R. (eds.) Computer Information Systems and Industrial Management, 16th IFIP TC8 International Conference, CISIM 2017, Białystok, Poland, Proceedings, pp. 131–140, 16–18 June 2017
13. Szymkowski, M., Saeed, E., Saeed, K.: Retina tomography and optical coherence tomography in eye diagnostic system. In: Chaki, R., Cortesi, A., Saeed, K., Chaki, N. (eds.) Advanced Computing and Systems for Security, vol. 5, pp. 31–42 (2018)
14. Szymkowski, M., Saeed, K.: Finger veins feature extraction algorithm based on image processing methods. In: Saeed, K., Homenda, W. (eds.) Computer Information Systems and Industrial Management, 17th International Conference, CISIM 2018, Olomouc, Czech Republic, 27–29 Sept 2018, Proceedings, pp. 80–92 (2018)
15. Bangare, S., Dubal, A., Bangare, P.S., Patil, S.: Reviewing Otsu's method for image thresholding. Int. J. Appl. Eng. Res. **10**(9), 21777–21783 (2015)
16. Saeed, K., Rybnik, M., Tabędzki, M.: Implementation and advanced results on the non-interrupted skeletonization algorithm. In: International conference on computer analysis of images and patterns, CAIP 2001, Warsaw, Proceedings, pp. 601–609 (2001)
17. Saeed, K., Tabędzki, M., Rybnik, M., Adamski, M.: K3 M: a universal algorithm for image skeletonization and a review of thinning techniques. Int. J. Appl. Math. Comput. Sci. **20**(2), 317–335 (2010)
18. Chen, W., Sui, L., Xu, Z., Lang, Y.: Improved Zhang-Suen thinning algorithm in binary line drawing applications. In: 2012 IEEE International conference on systems and informatics, ICSAI 2012, Yantai, China, Proceedings, 19–20 May 2012

19. Quinlan, J.R.: Induction of decision trees. Mach. Learn. **1**, 81–106 (1986)
20. Ren, J., Lee, S.D., Chen, X., Kao, B., Cheng, R., Cheung, D.: Naïve Bayes classification of uncertain data. In: 2009 IEEE 9th International Conference on Data Mining, Miami, USA, 6–9 Dec 2009, Proceedings

Biometric Fusion System Using Face and Voice Recognition

A Comparison Approach: Biometric Fusion System Using Face and Voice Characteristics

Aleksander Kuśmierczyk, Martyna Sławińska, Kornel Żaba and Khalid Saeed

Abstract This paper presents a biometric fusion system for human recognition which utilizes voice and face biometric features. The aim of the system is to verify or identify the users by their face and/or voice characteristics. The system allows reliable human recognition based on physiological biometric feature—face, and physiological–behavioral biometric feature—voice. The outcome yields better identification results using voice biometrics and better verification results using face biometrics. The approach used here for face recognition consists of an image processing being histogram equalization and application of Gabor filter, and extracting a feature vector from the image. Such a feature extraction method has been successfully applied in the iris recognition [3]. The speech recognition process uses a recording divided into frames out of which only the frames containing speech are considered. The feature vector is extracted from Mel frequency cepstral coefficient. The classification process is performed by a basic dynamic time warping algorithm. In this paper, evaluation of the above approaches for face and voice recognition is performed.

Keywords Face recognition · Text-dependent voice recognition · Biometrics · Biometric fusion system · Gabor filter · Mel frequency cepstral coefficients

A. Kuśmierczyk · M. Sławińska · K. Żaba (✉)
Faculty of Mathematics and Information Science,
Warsaw University of Technology, Warsaw, Poland
e-mail: kornelzaba@gmail.com

A. Kuśmierczyk
e-mail: aleksander.kusmierczyk@gmail.com

M. Sławińska
e-mail: martyna.slaw@gmail.com

K. Saeed
Faculty of Computer Science, Bialystok University of Technology, Bialystok, Poland
e-mail: k.saeed@pb.edu.pl

© Springer Nature Singapore Pte Ltd. 2020
R. Chaki et al. (eds.), *Advanced Computing and Systems for Security*,
Advances in Intelligent Systems and Computing 996,
https://doi.org/10.1007/978-981-13-8969-6_5

71

1 Introduction

The aim of this project is to construct a working biometric fusion system using face and voice features for human identification and verification, and to report its results on the independently collected dataset. The face feature vector consists of Gabor filter magnitudes and the color histogram moments for each color channel. The speech feature vector is built using static Mel frequency cepstral coefficients obtained from power spectrum of the voice signal in the Mel frequency scale. The identification and verification are done using minimum distance classifiers and dynamic time warping for face and speech, respectively.

In this project, there are two biometrics used, as such composition increases the possible success rate of human recognition. One of them is face which is a physiological biometric, and the second one is voice which is considered both physiological and behavioral biometrics. Physiological characteristics are set upon birth and rarely change throughout the lifetime. Behavioral characteristics are related to the behavior pattern of a person, which can change. The choice of these biometric identifiers was based on the ease of acquisition of the data, as both can be captured using one hardware device and the process of acquisition is not invasive. We use the dataset collected by our group on the university campus. The count of gathered samples is over a hundred. Most of the people who helped us collect the samples are Warsaw University of Technology's students. The dataset comprises the samples of people aged 20–30 years.

2 State of the Art

2.1 Facial Recognition Methods

The eigenvectors, which form an eigenimage, are derived from the covariance matrix of the probability distribution over the high-dimensional vector space of face images. "The approach transforms face images into a small set of characteristic feature images, called "eigenfaces," which are the principal components of the initial training set of face images." The results of using this method achieved the following success rates: 96% with light variation, 85% with orientation variation and 64% with size variation. The use of eigenfaces reduces the statistical complexity in face image representation and supports the use of large datasets. Furthermore, once the eigenfaces are calculated, they can achieve face recognition in real time. However, this method has significant drawbacks, and it requires highly controlled environment and has difficulty capturing expression changes [13].

Facial recognition using long-wave thermal infrared (LWIR)—research conducted by Diego Socolinsky and Andrea Selinger—uses the thermal images combined with regular visual cameras in order to achieve better performance than any of these systems separately. "Outdoor recognition performance is worse for both modalities, with a sharper degradation for visible imagery regardless of algorithm. It is clear from our experiments that face recognition outdoors with visible imagery are far less accurate than when performed under fairly controlled indoor conditions. For outdoor use, thermal imaging provides us with a considerable performance boost." The results of this study are as follows: Visual camera has the success rate of 97.05%, and LWIR has the success rate of 93.93%, while the fusion of the both reached 98.40% for the indoor performance. A drawback concerning the study is that the database used by the researchers was limited [8].

Nearest feature line (NFL) method is an extension of the principal component analysis (PCA) method. The data is processed using PCA in order to reduce its dimensionality. "The basic assumption is that at least two distinct prototype feature points are available for each class, which is usually satisfied. In a feature space, which is an eigenface space in this study, the NFL method uses a linear model to interpolate and extrapolate each pair of prototype feature points belonging to the same class" [12]. The average accuracy using this method on the dataset comprising of the pictures with changing orientation is as follows: Average accuracy is 80.01%, the highest achieved result is 100%, and the lowest is 24.44%. This method achieves the error rate lower by 34.6–56.4% with regard to the standard eigenface method [11].

2.2   Voice Recognition Methods

"Hidden Markov models (HMMs) have been used prominently and successfully in speech recognition." Hidden Markov models can be represented as the dynamic Bayesian network which is a set of variables related over two adjacent time steps (at any point in time, the value of a variable can be calculated from the preceding time point and the internal regressor). HMM is used to calculate vectors of cepstral coefficients of a short time frame. Such approach is often used in the deep learning software with the combination of Gaussian mixture model (GMM). The methods combined are often called GMM-HMM. The use of this method results in the following error rates depending on the hours of training and the training data: 36.2% of search error rate for voice searches as the training data and 24–48 h of training, and 21.7% of word error rate for switchboard (recordings of telephone calls) as the dataset and 2000 h of training [9, 10].

In contrast to phonetic-based approaches (HMM), which require separate components and training for its acoustic, language and pronunciation modes, the end-to-end automatic speech recognition (ASR) learns all the components of speech recognizer jointly, thus simplifying the training process as well as the deployment process. An attention-based model called "Listen, Attend and Spell" (LAS), further extended to "Watch, Listen, Attend and Spell" (WLAS) which is an ASR approach, is being developed and according to the researchers surpasses the human-level performance for speech recognition. LAS is an "end-to-end speech recognition model, no conditional independence, Markovian assumptions or proxy problems, and one model integrates all traditional components of an ASR system into one model (acoustic, pronunciation, language, etc.)."

Furthermore, ASR surpasses the previous attempts of its predecessor based on connectionist temporal classification (CTC) by not using conditional independence assumptions, thus not requiring the language model during the deployment, making it practical to use in devices with limited memory. The word error rate achieved by this solution was equal to 10.4% in the recent study for (LAS). The addition of lipreading module reduced this number to 7.9% [2, 4].

3   Algorithms

3.1   Voice Algorithm Description

The first step is to divide recording into the frames.

The stream of sampled voice recording is divided into frames of equal length. Length of frame in samples is obtained using Eq. (1).

$$F_{\text{len}} = t_{\text{len}} \times S_r t \, (\text{len}) \tag{1}$$

F_{len} is the frame length in seconds, and S_r is the sample rate of recording interval between frames obtained using Eq. (2).

$$F_{\text{int}} = t_{\text{int}} \times S_r t \, (\text{int}) \tag{2}$$

F_{int} is the frame interval in seconds, and S_r is the sample rate of recording.

As the last step of the preprocessing, the Hamming window is applied to each of the frames.

Hamming window goes over each frame of the audio recording, amplifies the middle of the frame and weakens the beginning and end of the frame as given in Eq. (3).

$$\omega(n) = \alpha - \beta * \cos\left(\frac{2\pi n}{N - 1}\right) \tag{3}$$

In Eq. (3), n is a sample in a frame, and N is a number of samples in a frame $\alpha = 0.53836$ and $\beta = 0.46164$.

Figure 1 presents the flowchart of extraction of speech feature vector.

The process of extracting the feature vector was achieved by fast Fourier transform (FFT) with Eq. (4)

$$x_k = \sum_{n=0}^{N-1} x_n \times e^{-i 2\pi kn/N} = \sum_{n=0}^{N-1} x_n \times [\cos(2\pi kn/N) - i \times \sin(2\pi kn/N)] \tag{4}$$

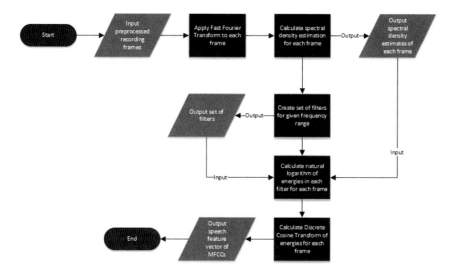

Fig. 1 Flowchart of speech feature extraction algorithm

Next, the spectral density estimation is calculated for each of the frames.

FFT is applied to an long frame, mapping the signal from time to spectral domain, resulting in n complex coefficients. Afterward, spectral density estimates, which describe strength of signal given the frequency, are calculated from first half of FFT coefficients.

Spectral density estimation of a single coefficient is given according to Eq. (5).

$$P(c_i) = \frac{|c_i|^2}{N} c_i \tag{5}$$

$P(c_i)$ is the ith coefficient of FFT, and N is the number of samples in frame.

Next step is to create set of filters for given frequency range and calculate the logarithm (natural or base 10) of energies for each filter in each frame.

Mel filter bank computes a set of Mel frequency cepstral coefficients (MFCCs) for a generated set of filters and spectral density coefficients of given frame.

Generation of filters is presented below.

For a given frequency range in Mel scale and n number of filters, $n + 2$ linearly spaced filter intervals are generated.

Filter intervals are converted back to Hertz scale.

Corresponding to the number of coefficients of FFT, $n + 2$ filter intervals are generated.

$$F_i = \left| \frac{(2 \times L + 1) \times f_i}{S_r} \right| \tag{6}$$

In Eq. (6), L is the number of coefficients of FFT, f_i—ith filter interval in Hertz scale and S_r—sample rate of the recording.

Finally, a filter bank of n vectors of triangular filters is computed.

Formula for obtaining kth filter is shown in Eq. (7) [6, 15].

For i from 0 to max(F) where F is set of ordered filter intervals:

$$\text{Filter}_k(i) = \begin{cases} 0, & i < F_{k-1} \text{ or } i > F_{k+1} \\ \frac{i-F_{k-1}}{F_k-F_{k-1}}, & F_{k-1} \le i \le F_k \\ \frac{F_{k+1}-i}{F_{k+1}-F_k}, & F_k \le i \le F_{k+1} \end{cases} \tag{7}$$

Generation of MFCC is presented below.

For each filter in a filter bank, the energy in the filter is obtained by the dot product between a filter vector and a vector of spectral estimates of a single frame.

Because humans perceive loudness in logarithmic scale, logarithm of filter energies is taken (8). The equation presents a spectral density estimate vector.

$$E_k = \ln(\text{Filter}_k \cdot P_{\text{frame}}) P_{\text{frame}} \tag{8}$$

E_k is the energy of kth filter.

Finally, a discrete cosine transform (DCT) of energies of a single frame is computed, resulting in a set of n real-valued Mel frequency cepstral coefficients [1, 3].

The last step of feature extraction is to calculate the DCT of energies for each frame.

Steps of extracting voice feature vector are enumerated below:

- Framing of the signal:
  A set of equally spaced intersecting frames is obtained from a vector of recording samples.
- Windowing the signal:
  In preparation for FFT, to each frame a window (Gauss or Hamming) function is applied.
- Obtaining spectral density estimates:
  Each frame is subjected to FFT, and a set of spectral density estimates is obtained.
- Creation of a Mel filter bank:
  For given length of FFT and sample rate, a filter bank of triangular filters is created. For each frame, its set of spectral density estimates is filtered through the filter bank, obtaining a set of energies left in each filter bank.
  A DCT of logarithms of energies is computed, resulting in a set of MFCC.
- Combining the MFCC sets:
  Sets of MFCC of each frame are concatenated, resulting in a feature vector of the recording.

The speech identification process uses dynamic time warping method. A flowchart of speech identification is presented in Fig. 2.

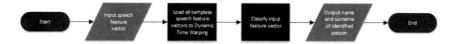

Fig. 2 Flowchart of speech identification algorithm

Dynamic time warping [4] is used for comparing two sequences of elements that differ in length. It returns the sum of distances between elements of those two sequences.

The function takes the input feature vector and checks the distance between this input vector and all the feature vectors stored in the database. After the calculation, the feature vector with smallest distance to the input feature vector is chosen as an answer.

The speech verification process uses dynamic time warping method described above. A flowchart of speech verification is shown in Fig. 3.

The function takes the input feature vector and the feature vector from the database that is suspected to be the same person's feature vector. The distance is calculated and checked whether it is within the set threshold. If the distance is smaller than the threshold, the answer is positive; otherwise, the answer is negative.

The process of training the dynamic time warping requires some number of speech feature vectors of the same person. The distances between each pair of feature vectors are calculated, and the feature vector with smallest average distance is selected. Next, it is stored in the database.

A general flowchart of training dynamic time warping algorithm is visible in Fig. 4.

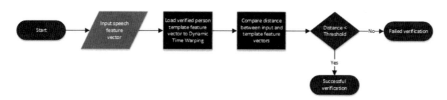

Fig. 3 Flowchart of speech verification algorithm

Fig. 4 Flowchart of training of dynamic time warping

3.2 Face Algorithm Description

The preprocessing phase of the face image includes equalizing the histogram of the image.

The process of histogram equalization works is presented in Fig. 5.

Histogram equalization calculates the cumulative distribution function (CDF) for the color channel values, which is also the image's accumulated normalized histogram and applies to the picture. Such operation increases the global contrast for the image, as the intensities of the color level are better distributed on the histogram.

CDF is calculated in Eq. (9).

$$CDF_x(i) = \sum_{j=0}^{i} p_x(j) \tag{9}$$

Variable $p_x(i)$ is the probability of an occurrence of a pixel of level i in the image and is calculated with the use of Eq. (10).

$$p_x(i) = p(x = i) = \frac{n_i}{n}, 0 \leq i \leq L \tag{10}$$

L is the number of color levels in the channel, n is the total number of pixels in the image, and n_i is the number of occurrences of the color level i.

In order to apply this method back to the original picture, the final function is used given in Eq. (11).

$$y = \lfloor CDF(i) \times (((max(x) - min(x)) + min(x))) \rfloor \tag{11}$$

The resultant value y is the new pixel color level for the channel x from the original color level i.

Figure 6 shows a flowchart of face feature extraction.

The face feature vector consists of two feature vectors:

- Color histogram-based feature vector:

First, the color histogram is calculated for a given bitmap and stored in three arrays (one array per one channel). As a next step, those color histograms are used to calculate statistical moments per channel.

The process of calculating the moments is the following:

Color moment generating is generated using Eqs. (12) and (13) [2].

Fig. 5 Flowchart of histogram equalization

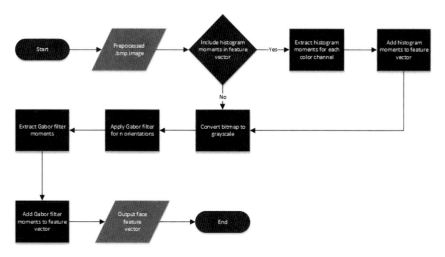

Fig. 6 Flowchart of face feature extraction

$$M_k^1 = \frac{1}{XY} \sum_{x=1}^{X} \sum_{y=1}^{Y} f_k(x, y) \tag{12}$$

For the first order,

$$M_k^h = \left(\frac{1}{XY} \sum_{x=1}^{X} \sum_{y=1}^{Y} (f_k(x, y) - M_k^1)^h \right)^{\frac{1}{h}} \tag{13}$$

k represents the kth color component.
h represents order of the moment.
$f_k(x, y)$ represents the color value of the kth color component in the pixel at the coordinates of (x, y).

The moments generated by the aforementioned functions are mean for the first order, variance for the second order and skewness for the third order, and all efficiently describe the color distributions of images.

- Gabor filter magnitude-based feature vector:

In order to calculate Gabor filter magnitudes, the image must be converted into the gray scale. The process of conversion is described below.

To transform the image into grayscale image, the function normalizes loops through every pixel of the image and sets the value for each channel of given pixel to the average value of values of all channels of that pixel.

The pseudocode is the following (actions done for each pixel):

channelValue = (pixel.Red + pixel.Green + pixel.Blue)/3

pixel.Red = channelValue
pixel.Green = channelValue
pixel.Blue = channelValue

Next, the Gabor filter is applied to the image, and the filter is described below.

Gabor filter is an edge detection filter. The edges of an image are detected from different orientations, and next they are placed on top of one another to get the final image with detected edges.

Equation (14) is used to calculate Gabor filter [5].

$$G_{\sigma_y,\sigma_x,f_i,\theta_k}(x,y) = e^{-\left|\frac{x_{\theta_k}^2}{\sigma_x^2} + \frac{y_{\theta_k}^2}{\sigma_y^2}\right|} \times \cos\left(2\pi f_{ix\theta_k} + \varphi\right)$$

$$x_{\theta_k} = x \times \cos\theta_k + y \times \sin\theta_k$$

$$y_{\theta_k} = y \times \cos\theta_k - x \times \sin\theta_k$$

$$\varphi = \pi/2$$

$$\theta_k = \frac{k\pi}{n} \tag{14}$$

The next step is to calculate the kernel by applying the formula given in Eq. (15) to elements of the kernel.

$$\text{kernel}[x,y] = e^{-\left|\frac{x_{\theta_k}^2}{\sigma_x^2} + \frac{y_{\theta_k}^2}{\sigma_y^2}\right|} \times \cos\left(2\pi \frac{x_{\theta_k}}{\text{waveLength}} + \varphi\right) \tag{15}$$

The last step is to calculate the Gabor filter magnitudes. Equations (16), (17) and (18) are used in the calculations. They represent the magnitudes (mean, standard deviation and skewness) [2].

$$\mu = \frac{1}{XY} \times \sum_{x=1}^{X}\sum_{y=1}^{Y} \text{image After Gabor Filtering}(x,y) \tag{16}$$

$$\text{std} = \sqrt{\frac{1}{XY} \times \sum_{x=1}^{X}\sum_{y=1}^{Y}||\text{image After Gabor Filtering}(x,y)| - \mu|^2} \tag{17}$$

$$\text{skew} = \frac{1}{XY} \times \sum_{x=1}^{X}\sum_{y=1}^{Y}\left(\frac{\text{image After Gabor Filtering}(x,y) - \mu}{\text{std}}\right)^3 \tag{18}$$

The above two feature vectors are combined into one face feature vector.

The face identification process uses minimum distance classifier method [7, 16]. A flowchart of face identification algorithm is presented in Fig. 7.

This method analyzes the numerical face feature vectors and calculates the Euclidean distance between them in order to further use the calculated distance in determining how similar the given vectors are.

Fig. 7 Flowchart of face identification algorithm

Euclidean distance is calculated using Eq. (19).

$$d(p, q) = \sqrt{(q_1 - p_1)^2 + (q_2 - p_2)^2 + \cdots + (q_n - p_n)^2} \qquad (19)$$

The function takes the input feature vector and checks the distance between this input vector and all the feature vectors stored in the database. After the calculation, the feature vector with smallest distance to the input feature vector is chosen as an answer.

The face verification process uses minimum distance classifier method described above. Figure 8 shows a flowchart of face verification process.

The function takes the input feature vector and the feature vector from the database that is suspected to be the same person's feature vector. The distance is calculated and checked whether it is within the set threshold. If the distance is smaller than the threshold, the answer is positive; otherwise, the answer is negative.

The process of training the minimum distance classifier requires some number of face feature vectors of the same person. Those feature vectors are used to calculate the average feature vector for that person, which is stored in the database.

In Fig. 9, a flowchart of minimum distance classifier training is presented.

Fig. 8 Flowchart of face verification algorithm

Fig. 9 Flowchart of training of minimum distance classifier

4 Dataset Description

The datasets used in testing the algorithms have been personally gathered by the authors of the thesis in order to obtain datasets with consistent external factors such as lighting and noise. Datasets are composed of recordings of words "close" and "algorithm," and photographs of faces.

4.1 Used Equipment and Format of Data

All photographs and recordings were taken with camera Logitech HD Webcam C270 with an embedded microphone. Process of recording was facilitated by using FFmpeg audio/video processing library [14].

Each photograph of person's face was formatted as a 24-bit color bitmap with resolution 640 × 480 (width × height) pixels.

Each recording of spoken word was formatted as signed 16-bit PCM WAV file with sample rate 44,100 Hz and length of 3 s.

From each face image, redundant background was cropped and from each voice recording, redundant silence was cut off.

Bitmaps and WAV files containing huge loss of data (person not present in an image, voice recording cut short) were discarded.

4.2 Acquisition of the Datasets

The datasets gathering sessions took place from November 2017 till January 2018. Samples were gathered in a room of MINI faculty of Warsaw University of Technology. The chosen room was artificially illuminated and mostly secure from the noise and echo.

Recorded persons were invited one by one. From each of them, six photographs of face, six recordings of word "algorithm" and six recordings of word "close" were collected. Each person was situated at distance of approximately 0.75 m in front of the camera, with eyes positioned at the same level as the lens of the camera.

All photographs were taken in quick succession, with allowance of minor changes of the face expression. Each recording was made after a prompt from the recording person, with allowance of minor changes of pitch and speed of speech.

Table 1 Dataset information

Name	Face training photographs	Face test photographs	Algorithm training recordings	Algorithm test recordings	Close training recordings	Close test recordings	Total size
Data	324	124	326	123	328	126	81/82

4.3 Division of the Datasets

Datasets comprise the specified amount of records (one record is the set of 6 photographs and 12 recordings per person) visible in Table 1. Due to data corruption encountered during acquisition process, the dataset contains one record lacking the facial imagery.

5 Test Results

Below presented are the results for identification of a person among the people stored in the database, and the parameters used for face and voice identification algorithms.

5.1 Voice Recognition Parameter Description

In Fig. 10, a spectrogram of a recording in spectral domain is presented.

Fig. 10 Spectrogram of a voice recording, where *x*-axis corresponds to time, *y*-axis to frequency and color spectrum to amplitude

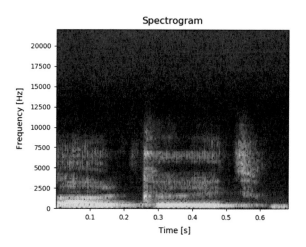

Fig. 11 Triangular filter
bank consisting of 11 filters
for frequencies in 0–8000 Hz
range

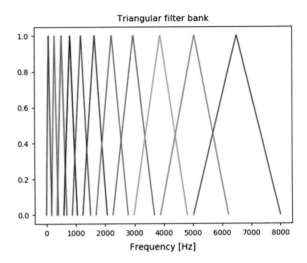

We have chosen five parameters to be adjusted for voice recognition. By the method of trial and error, we have chosen the values of parameters for best quality of voice feature extraction (presented within brackets).

- Length of a single frame of the recording (0.03 s),
- Interval between each frame of the recording (0.015 s),
- Number of filter banks used in calculation of MFCCs (18 filter banks),
- Number of Mel frequency cepstral coefficients taken from extraction (11 coefficients),
- Type of a window function applied to the recording samples (Hamming window function),
- Base of logarithm of filter bank energies (e or 10).

In Fig. 11, a visualization of a triangular filter bank is presented.

5.2 Face Recognition Parameter Description

We have chosen four parameters to be adjusted for face recognition. By the method of trial and error, we have chosen the values of parameters for best quality of face feature extraction using both color histograms and Gabor filter (presented within brackets for both color histograms and Gabor filter). In Fig. 12 are results of histogram equalization applied to a picture, which is used in color histogram moment calculation.

- Input wavelength of Gabor filter (4 pixels),
- Number of orientations (angles) of Gabor filter (7),
- Horizontal standard deviation of Gabor filter (6),

Fig. 12 Face image before and after histogram equalization

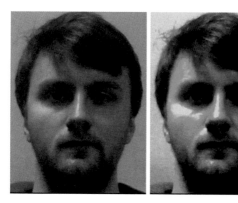

– Vertical standard deviation of Gabor filter (7.5).

By the method of trial and error, we have chosen following values of parameters for best quality of face feature extraction using only Gabor filter magnitudes:

– Input wavelength of Gabor filter: 1 pixel,
– Number of orientations (angles) of Gabor filter: 7,
– Horizontal standard deviation of Gabor filter: 1,
– Vertical standard deviation of Gabor filter: 1.

In Fig. 13 are the steps of application of the Gabor filter resulting in the final image used for Gabor filter magnitude calculation.

The identification and verification test results for speech and face recognition of the dataset consisting of over 80 persons are presented in Table 2.

6 Conclusions

During our work, we reached the following conclusions.

Results of the face identification process, using Gabor magnitudes and histogram magnitudes, signify that the combination of the aforementioned is the most effective from the chosen methods. Furthermore, low success rate of histogram magnitudes indicates that it is not usable as a singular extraction method for face recognition.

From the results of voice verification tests, where the same verification threshold was assigned to each word, we observe that with the increase of the dataset size, the FRR decreases for both words, while the FAR increases; furthermore, we have determined the optimal values for the thresholds for the datasets.

We observe a higher success rate for the monosyllabic word in comparison with the multisyllabic. We assume that this conclusion can have multiple origins. First, the word chosen—"algorithm"—is considered as hard to pronounce for non-native English speakers, who are the main source of samples in the dataset. Thus, the quality of recordings is worse than in the case of the monosyllabic word—"close".

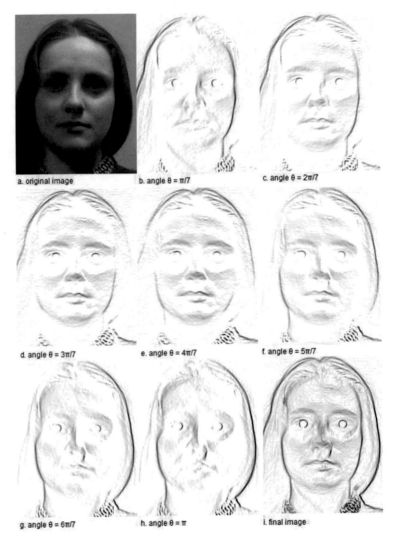

Fig. 13 Face image before and after the use of Gabor filter for different values of angle θ

Additionally, the recordings of "close" are on average louder, hence higher success rate of the aforementioned word.

The study shows that voice recognition is more successful for identification than verification and face recognition is a more viable mean of verification than identification. Furthermore, the study shows that all of the above conclusions are drawn for the dataset, which in the field of biometrics is considered to be too small, thus inconclusive for most of the research. However, the dataset allows us to test the general viability of the methods used in the project.

Table 2 Results of identification and verification

Test	Success rate in % (%)	Success rate of test samples	FAR in % (%)	FAR of test samples	FRR in % (%)	FRR of test samples	Number of tested persons	Number of persons in dataset
"Algorithm" speech identification	63.41	78/123					63	82
"Close" speech identification	71.43	90/126					63	82
"Algorithm" speech verification			17.42	1736/9963	18.70	23/123	63	82
"Close" speech verification			36.52	3727/10,206	11.90	15/126	63	82
Gabor + color face identification	45.16	56/124					62	81
Gabor only face identification	44.35	55/124					62	81
Color only face identification	21.77	27/124					62	81
Gabor + color face verification			22.50	2232/9920	6.45	8/124	62	81

(continued)

Table 2 (continued)

Test	Success rate in % (%)	Success rate of test samples	FAR in % (%)	FAR of test samples	FRR in % (%)	FRR of test samples	Number of tested persons	Number of persons in dataset
Gabor only face verification			22.81	2263/9920	4.84	6/124	62	81
Color only face verification			24.68	2448/9920	9.68	12/124	62	81

The overall results of the project are promising as the unusual methods have been tested and compared. Considering the size of the sample, we have deduced that our recognition system may be enhanced by switching the voice classifier to a state-of-the-art algorithm such as hidden Markov model or recurrent neural network. Furthermore, face recognition accuracy may be improved by extracting additional geometric face features and Haar-like features.

Bibliography

References

1. Chakraborty, K., Talele, A., Upadhya, S.: Voice recognition using MFCC algorithm. Int. J. Innov. Res. Adv. Eng. (IJIRAE) **1**(10) (2014). ISSN: 2349-2163
2. Chan, W., Jaitly, N., Le, Q., Vinyals, O.: Listen, attend and spell: a neural network for large vocabulary conversational speech recognition. In: ICASSP (2016)
3. Choraś, R.: Image feature extraction techniques and their applications for CBIR and biometrics systems. Int. J. Biol. Biomed Eng. **1**(1) (2007)
4. Chung, J.S., Senior, A., Vinyals, O., Zisserman, A.: Lip reading sentences in the wild (2016). arXiv:1611.05358
5. Furui, S.: Digital Speech Processing Synthesis and Recognition, 2 edn. (2001)
6. Muda, L., Begam, M., Elamvazuthi, I.: Voice recognition algorithms using Mel Frequency Cepstral Coefficient (MFCC) and Dynamic Time Warping (DTW) techniques. J. Comput. **2**(3). ISSN 2151-9617
7. Rouhi, R., Amiri, M., Irannejad, B.: A review on feature extraction techniques in face recognition. Sig. Image Process. Int. J. (SIPIJ) **3**(6) (2012)
8. Socolinsky, D.A., Selinger, A.: Thermal face recognition in an operational scenario. In: IEEE Computer Society, pp. 1012–1019 (via ACM Digital Library) (2004)
9. Starner, T., Pentland, A.: Real-time American sign language visual recognition from video using hidden Markov models. Master's thesis, MIT Program in Media Arts (1995)
10. Deng, L., Yu, D.: Deep learning: methods and applications. Found. Trends Sig. Process. **7**(3–4), 197–387 (2014)
11. Swaminathan, A.: Face recognition using support vector machines. In: ENEE633: Statistical and Neural Pattern Recognition (2005)
12. Li, S.Z., Lu, J.: Face recognition using nearest feature line method. IEEE Trans. Neural Netw. **10**(2), 439–443 (1999)
13. Turk, M., Pentland, A.: Face recognition using eigenfaces. In: Computer Vision and Pattern Recognition. Proceedings CVPR'91. IEEE Computer Society Conference (1991)

Internet Sources

14. FFmpeg official website. https://www.ffmpeg.org/. Accessed 6 Dec 2017
15. Mel Frequency Cepstral Coefficient (MFCC) tutorial. http://practicalcryptography.com/miscellaneous/machine-learning/guide-mel-frequency-cepstral-coefficients-mfccs/. Accessed 25 Nov 2017
16. Minimum Distance Classifiers tutorial. https://homepages.inf.ed.ac.uk/rbf/HIPR2/classify.htm. Accessed 6 Dec 2017

Pattern Recognition and Imaging

A Multi-class Image Classifier for Assisting in Tumor Detection of Brain Using Deep Convolutional Neural Network

Abhishek Bal, Minakshi Banerjee, Punit Sharma and Rituparna Chaki

Abstract Segmentation of brain tumor is a very crucial task from the medical points of view, such as in surgery and treatment planning. The tumor can be noticeable at any region of the brain with various size and shape due to its nature, that makes the segmentation task more difficult. In this present work, we propose a patch-based automated segmentation of brain tumor using a deep convolutional neural network with small convolutional kernels and leaky rectifier linear units (LReLU) as an activation function. Present work efficiently segments multi-modalities magnetic resonance (MR) brain images into normal and tumor tissues. The presence of small convolutional kernels allow more layers to form a deeper architecture and less number of the kernel weights in each layer during training. Leaky rectifier linear unit (LReLU) solves the problem of rectifier linear unit (ReLU) and increases the speed of the training process. The present work can deal with both high- and low-grade tumor regions on MR images. BraTS 2015 dataset has been used in the present work as a standard benchmark dataset. The presented network takes T1, T2, T1c, and FLAIR MR images from each subject as inputs and produces the segmented labels as outputs. It is experimentally observed that the present work has obtained promising results than the existing algorithms depending on the ground truth.

Keywords Segmentation · Deep learning · Brain tumor · Deep convolutional neural network · MRI

A. Bal (✉) · M. Banerjee
RCC Institute of Information Technology, Kolkata, India
e-mail: abhisheknew1991@gmail.com

M. Banerjee
e-mail: mbanerjee23@gmail.com

P. Sharma
Apollo Gleneagles Hospital, Kolkata, India
email: dr_punitsharma@yahoo.com

R. Chaki
A. K. Choudhury School of Information Technology,
University of Calcutta, Kolkata, India
e-mail: rituchaki@gmail.com

© Springer Nature Singapore Pte Ltd. 2020
R. Chaki et al. (eds.), *Advanced Computing and Systems for Security*,
Advances in Intelligent Systems and Computing 996,
https://doi.org/10.1007/978-981-13-8969-6_6

1 Introduction

Among the various types of brain tumor [1], gliomas is the most common one, which is sub-categorized into less aggressive and highly aggressive [1, 2]. A patient with less aggressive tumor can live for many years, whereas the life expectancy of highly aggressive tumor patients is at most two (2) years. Thus, accurate segmentation and delineation are two crucial tasks for better diagnosis [1, 2] that can improve the quality of patient's life. The common treatment plan to deal with brain tumor is surgery, but in some critical cases, where surgery is not possible due to various constraints, alternative techniques such as radiation can decrease tumor growth. The structural details of the human brain are captured by MRI, which contributes significantly to the diagnosis of brain tumor, such as growth prediction and in treatment planning. Deep knowledge about the brain tissues is needed during the diagnosis of tumor.

Tumor can occur at any region of the brain with various sizes and shapes due to its nature, which makes the segmentation process more difficult. MRI-based brain tumor segmentation [3, 4] also depends on the qualities of the MR machine and the acquisition techniques. If these two parameters are varied, then the intensity value of same tumorous cell may also differ in different images. Brain tissues are normally divided into three parts, namely white matter, gray matter, and cerebrospinal fluid (CSF). The focus of the tumor analysis of the brain is to identify the tumorous tissue, such as enhance tumor, non-enhance tumor, necrosis, and edema. It is seen that the boundary regions of the tumor are fuzzy in nature, which is very difficult to distinguish from the normal tissues. To deal with such type of problem, multi-modalities MR images are captured during the acquisition process, such as FLAIR, T1, T1c, and T2. The combination of these modalities provides a better outcome than a single modality.

Almost all previous methods except the deep neural network worked on hand-crafted features, which may require the advance knowledge on the datasets. The handcrafted features problem can be removed by the concept of deep learning, where features [5] are automatically extracted depending on the natures of datasets. The most popular method of deep learning is convolutional neural network (CNN), that works significantly well in image recognition, object recognition, and speech recognition. The raw data can be directly used as the inputs of CNN, where inputs are convolved with a set of learnable filters to compute the output feature maps. The values of the filters are called weights. The features map convoluted with the share-able filters or the kernels. The weight sharing in deep neural network improves the performance of the different applications over the traditional artificial neural network and also decreases the computational load in CNN architecture. Basically, the fully connected (FC) layers in the artificial neural network increase computational load due to the massive interconnection that causes overfitting. Regularization methods in the neural networks, such as dropout can reduce the overfitting problem by removing unwanted nodes depending on its probability and confirms to extract the best features [5, 6]. Several methods have been proposed [7–10] from the past few years based on deep convolutional neural networks. Inspired by the previous deep convolutional neural network based works, present work proposes an efficient patch-based CNN architecture for dealing with various types of brain tumor by introducing small convolutional kernels (3×3 filters), activation function such as leaky rectifier linear units (LReLU) and dropout [6] as regularization. The 3×3 filters allow to design a deep architecture (large number of convolution layers) by reducing the weights in each layer and also allows non-linearity to the dataset by increasing number of layers which makes a positive effect in removing the overfitting. In the proposed method, leaky rectifier linear unit (LReLU) is used as an activation function to overcome the problem of rectifier linear units (ReLU). Beside of these, present work also uses the rotation based (multiple of $90°$) data augmentation for increasing the amount of training data. Experimental analysis of the present work has been taken place on BraTS 2015 dataset, which is sub-categorized into training and challenge datasets. This brain tumor segmentation (BraTS 2015) datasets helps us to compare the present method with the other methods with respect to ground truth based on complete, core and enhanced tumor regions segmentation.

The remaining contents are described into five sections. Related work is briefly presented in Sect. 2. The proposed work with theoretical details are described in Sect. 3. The brief details of the several metrics are described in Sect. 4. The Sect. 5 presents experimental details of the proposed method. The conclusion is presented in Sect. 6.

2 Related Work

Rajendran et al. [11] proposed a method that integrates region-based fuzzy clustering and deformable model for MRI brain tumor segmentation. In their proposed method, an initial segmentation of the tumor is performed with fuzzy clustering, and then the outcome is used for deformable models to determine the exact tumor region using external force such as gradient vector field. Menze et al. [1] presented a fusion approach of several tumor detection algorithms which is applied on 65 multi-contrast MR images. It performs well than a single algorithm, because it uses hierarchy majority voting, which efficiently handle all tumor sub-regions. Hamamci et al. [12] introduced brain tumor segmentation process using cellular automata by incorporating a parameter that deals with heterogeneous tumor detection. They used contrast-enhanced T1-weighted MR image for verifying their proposed algorithm. Havaei et al. [13] proposed a brain tumor segmentation method that requires communication from the user. Havaie et al. [13] improved the performance of their proposed MRI brain tumor segmentation method by introducing some spatial feature of coordinates to intensity features. A 3D segmentation method of brain tumor using fuzzy clustering is proposed by Khotanlou et al. [14] that can deal with various types of tumor. The deformable model along with spatial relations is used during the segmentation of brain tumor. The outcomes of the segmentation were validated against the ground truth. Selvakumar et al. [15] introduced an automated brain tumor detection method by combination of K-means and fuzzy C-mean clustering method. They performed statistical calculations for measuring the exact position and the shape of the detected tumor region. Selvakumar et al. [15] shown that their method performs well with respect to manual segmentation. Rajendran et al. [16] presented a brain tumor detection method by applying GVF deformable model on the output of the region-based fuzzy clustering method. GVF deformable model helps to determine the contour of tumor boundary. Zikic et al. [17] proposed an architecture of convolutional neural networks (CNN) for segmentation of brain tumor using the small patches, which are extracted from the different MRI modalities. A standard intensity normalization method is used as a preprocessing step to restrict the variety of MR scanner. Their network achieved promising outcomes on BraTS 2013 challenge dataset. Havaei et al. [10] proposed an architecture of convolutional neural network to segment brain tumor. Their presented architecture can efficiently deal with high- and low-grade tumors using local and global features through cascade architecture. The experimental results of their proposed network achieved promising outcomes on BraTS 2013 test dataset than other methods. Hussain et al. [18] proposed a patch-based CNN architecture for multi-class classification of brain tumor. They used two different patches (neighborhoods) for each pixel such as 37×37 and 19×19. To reduce the over-fitting in their network, max-out and dropout are used. Rao et al. [19] proposed a CNN architecture which is trained separately using four MRI modalities by separating the patches from three different planes (axial, coronal, and sagittal). The outcomes of the four different CNN trained networks are concatenated and used as an input feature set in the random forest (RF), which finally classifies each pixel. In their CNN network [19], 5×5 convolutional kernels are used. Dvořák et al. [9] presented a local structure prediction approach-based convolutional neural network as a learning algorithm to segment tumor regions. Their network has been applied on BraTS 2014 dataset and achieved better performance than other existing methods. A 3D brain tumor segmentation technique for MR image has been reported by Wels et al. [20] which combined Markov random field with graph cut technique. The proposed discriminative method depends on observed local values and surrounding context of the tumor regions without any user interaction. A small convolutional kernel-based CNN architecture has been proposed by Pereira et al. [21] that can efficiently segment the brain tumor from MR image. Data augmentation and intensity normalization in their method have shown that preprocessing can increase the performance of the proposed method. Their network validated on BraTS 2013 challenges dataset. Kaus et al. [22] presented 3D segmentation method for assisting the brain tumor on MR data and verified their method using the ground truth. MR images of 20 patients have been used for their experiment. Their method takes very small amount of times with respect to manual segmentation.

Most of the deep learning-based brain tumor segmentation methods which are highlighted in this literature used large parameters that increase the complexity of the architecture. In order to reduce the parameter and better performances, present work proposes an efficient CNN architecture with small convolutional kernels (3×3 filters) and LReLU activation function that allows to design a deep architecture by reducing the number of weights in each layer and more nonlinearity on the datasets.

3 Proposed Method

The block diagram in Fig. 1 represents the steps of the present work. The three main steps of the present work are preprocessing (Sect. 3.1), CNN-based classification (Sect. 3.2), and post-processing (Sect. 3.3).

3.1 Pre-processing

The main difficulties of handling MR image is the intensity inhomogeneity. Due to the different constraints, like MR machine quality and capturing technique of MR data, the intensity inhomogeneity appears in MR image. Intensity inhomogeneity may lead to misclassification. So, the proposed method first removes intensity inhomogeneity from original MR image using bias field correction method before further processing. The bias field correction algorithm in Ref. [23] is used in our proposed method. The effect of bias correction is shown in Fig. 2. Next, tumor patches are extracted from axial slices, including four modalities. Finally, zero mean and unit deviation-based normalization has been applied in extracted patches. During the patch normalization, mean and standard deviation are calculated across all the patches for each modality separately. Finally, the mean value is subtracted across all the patches for each modality, which called zero-centered normalization, then

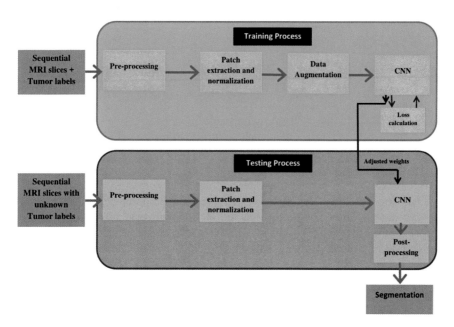

Fig. 1 Block diagram proposed model

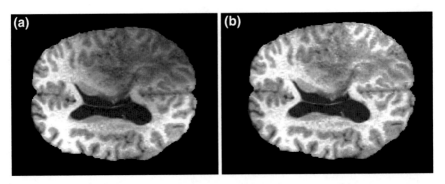

Fig. 2 Effect of bias correction [4]. **a** Original MRI, **b** bias field corrected MRI

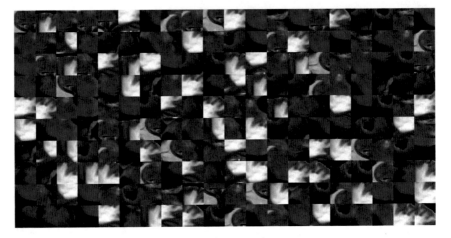

Fig. 3 Few 33 × 33 tumor patches with different classes in T2 MRI modality

the resulting patches are divided by the standard deviation for each modality. Few 33 × 33 patches (before zero mean and unit deviation) from T2 MRI modality are shown in Fig. 3.

Due to class imbalances in brain tumor images, very less amount of tumor tissues exist as compared to normal tissues. There are some classes in tumor regions that are more common than the other tumor tissues. It is seen that edema covers larger volume than necrosis. On the other hand, for few patients, necrosis may not exist. In our training datasets, 35% are extracted as normal tissues and remaining 65% as tumor tissues. Almost balanced number of tumor classes are chosen among the 65% tumor tissues. In tumor tissues, the occurrences of some classes are very rare, so for class balancing during training, artificial data augment is performed by rotating the patches with different angles (multiple of 90°). But during the testing phase, the present method kept the testing patches as it is without any data augmentation.

3.2 Convolutional Neural Network

In the present work, segmentation is carried out slice by slice on the axial plane by classifying each pixel by processing $Y \times Y$ patch, where the pixel to be classified is situated in the middle pixel position of the patch. The input dimension of the CNN architecture is represented by $Y \times Y \times 4 \times Z$, where Z represents the total patches number and 4 represents four different MRI modalities. In this architecture, several convolutional layers with activation functions, pooling layers, and fully connected layers are stacked together, which are shown in Fig. 4. Each of these layers is able to extract the important features of the preceding layer called feature maps.

Convolutional Layer Convolutional layers are the heart of the convolutional neural network. In the convolutional layer, kernels or filters are convoluted with convolutional layer's feature maps and produces the input (features maps) to the next layer. Kernels are basically a collection of weights, which are initialized randomly or through some initialization methods [24]. During training procedure, kernel weights are optimized through back-propagation to find out the essential features of the original image. The ith feature map (C_i^l) in lth layer is computed as:

$$C_i^l = b_i^l + \sum_j W_{ij}^l * X_j^{l-1} \tag{1}$$

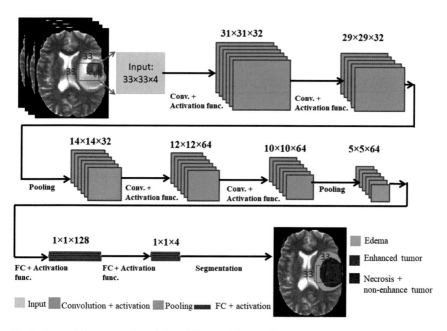

Fig. 4 Steps of the proposed patch-based CNN architecture for tumor segmentation on MR image

where b and W represent the bias and the weights of the kernel, respectively. X_j^{l-1} denotes previous layer's $(l-1)$ jth feature maps. The size of the kernels plays a very significant role during convolution. The kernel's size in the present architecture is 3×3. We choose small size kernel because more layer can be introduced to form a deeper architecture. For example, same size feature map can be produced by 3×3 kernels (stride 1) with two layers and 5×5 kernels (stride 1) with one layer. But the advantage of choosing 3×3 kernels is that less number of the kernel's weights are required during training. In the present work, 33×33 patches are chosen as input from each of the four MRI modalities.

In the present CNN architecture, the valid-mode operation is used during convolution. So, the convolution operation is performed only for the pixels which have minimum $K/2$ number of neighborhoods on each side. $Y \times Y$ input patch (feature map) is convoluted with $K \times K$ kernel and produces $F \times F$ output as a feature map, where $F = Y - Z + 1$. In our CNN architecture, K is always 3, that denotes same height and width of the kernels. So, if $Y = 11$, by $F = 9$.

Pooling Pooling layer performs the downsampling to the features map of the previous layer. During pooling, the target is to keep the important features in the features map and decreases the computational load, that makes easier for next layer computation. Several types of pooling are available such as max-pooling, mean-pooling, min-pooling, etc. Each type of pooling can be performed as overlapping or non-overlapping way. In the present method, we choose max-pooling with the non-overlapping concept that decreases the dimension of the feature maps while keeping the maximum features within the pooling sub-window (P).

The sub-window size depends on the application. Generally, small size sub-window is chosen to restrict the loss of important features. The sub-window may or may not be overlapped. The shrinking during max-pooling depends on the dimension of the pooling sub-window (P) with the stride parameter (S). The stride (S) parameter controls the positional increment of pooling sub-window toward the horizontal and vertical directions. The pooling operation is formulated as:

$$PF = (F - P)/S + 1 \tag{2}$$

In Eq. 2, $F \times F$ and $PF \times PF$ denote the feature map dimension before and after pooling. The pictorial presentation of max-pooling is shown in Fig. 4. The proposed work considers $P = 2$ and $S = 2$ (Fig. 5).

Fully Connected (FC) Layer Fully connected layer in CNN work as the similar way like in the artificial neural network. A fully connected layer can be represented as:

$$y = Wx + b \tag{3}$$

In Eq. 3, W represents the weight matrix. The input and output of the FC layer are represented by x and y, respectively. The bias vector is represented by b.

Activation Function The nonlinearity of the dataset is performed by the activation function. There are several activation functions available, namely sigmoid, tangent,

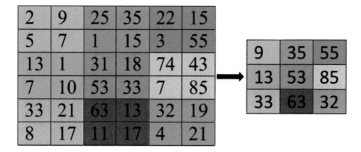

Fig. 5 Max-pooling operation with 2×2 pooling window and stride 2

rectifier linear units (ReLU), etc. Among the several types of activation functions, most of the deep neural network-based methods use the rectifier linear units (ReLU) for achieving better results, which is formulated as:

$$f(X) = \max(X, 0). \tag{4}$$

The major advantages of ReLU are sparsity and reduced likelihood of vanishing gradient. In ReLU, when $X > 0$, the gradient has a constant value. In ReLU, sparsity arises when $X \leq 0$, this makes the advantages in terms of saving some energy. Rectifier linear units (ReLU) also help to increase the speed of the training process [6, 25]. Besides these advantages, ReLU suffers from a problem called "dying." In Eq. 4, it is seen that a constant 0 can be initialized if the value of X is negative, that can affect the weights optimization during back-propagation [26]. That means, if the ReLU output is 0, then the gradient through it will also 0. So, during back-propagation, no error signal propagated from the later layers to earlier layers. To overcome this problem, a variation of rectifier linear units (ReLU) is introduced called leaky rectifier linear unit (LReLU) [26]. Little bit slope is added in the negative portion using the LReLU activation function by which the gradient is not able to reach to 0. It is formulated as:

$$f(X) = \max(X, 0) + \alpha \times \min(0, X). \tag{5}$$

In Eq. 5, α represents the leakiness parameter with value 0.33. In the present architecture, an element-wise LReLU is computed to the results of the convolutional layer and FC layer (except last FC layer) to obtain nonlinear feature maps. The number of neurons in last FC layer is same as total number of classes. Each neuron in the last layer detects one class properties. In the outcomes of the last layer (FC layer), softmax operation is used to normalize the final outcomes of the network (last FC layer's outcome), which is formulated as:

$$F(X_i) = \frac{\exp(X_i)}{\sum_{j=1}^{n} \exp(X_j)} \tag{6}$$

where X is the outcomes (feature maps) of the last FC layer and $F(X_i)$ denotes the final outcomes after applying the softmax activation. The present work uses cross-entropy as a loss function, which is formulated as:

$$\text{loss} = - \sum_{i \in \text{voxels}} \sum_{j \in \text{classes}} c_{ij} \log(o_{ij}) \tag{7}$$

where o and c denote the predicted outcome after softmax activation function and target, respectively. During training, loss function is optimized through stochastic gradient descent with the momentum coefficient of 0.9. In the present method, 0.01 is chosen as an initial learning rate (α_0). Here, the decay factor (DF) is 10^{-1}. The learning rate (α) is reduced after each epoch using the Eq. 8.

$$\alpha = \frac{1}{1 + \text{DF} \times \text{EN}} \times \alpha_0 \tag{8}$$

In Eq. 8, α_0 and EN denote the initial learning rate and current epoch number, respectively.

Regularization To minimize the overfitting in the FC layer, regularization is used. The present method implements the dropout [6, 27] to remove the overfitting in fully connected (FC) layers. During the training period, the dropout removes the nodes in FC layers which has the probability with p. This removal helps to learn better training network (better representation of data) and achieves significant performance on test datasets.

Data Augmentation It is used in the neural network to artificially increases the amount of the training dataset. It also removes the problem of overfitting [25] in the deep neural network. Several types of data augmentation are already proposed [28–30]. It is also seen that improper augmentation makes a negative impact on the training process and will result misclassification of the dataset. Since in the case of MRI, huge numbers of normal and abnormal datasets are present. For that reason, some of the deep learning [10, 31] based MRI brain tumor studies did not use the data augmentation. In our proposed method, it is observed that rotation-based data augmentation increases proposed network performance by using class balancing. Present work rotates the patches by multiple of 90°, namely, 90°, 180°, and 270°.

3.3 Post-processing

Sometimes small cluster regions are misclassified as abnormal or tumor tissues. To deal with such problem, post-processing is applied after CNN using connected component analysis. In the post-processing, present work removes the cluster regions (components) which are smaller than the threshold value.

4 Quantitative Analysis

The few quantitative indices that are used to measure the performance of the proposed method are briefly described in this section. Although present work segments each subject into four classes, namely normal tissues, edema, enhanced tumor, and non-enhanced tumor + necrosis. But quantitative analysis has taken place in three regions, namely complete tumor region (considering all tumor classes), core region (enhance tumor, non-enhance tumor, and necrosis), and enhance or active tumor region. Let, for each region, present work obtains a binary map $PR \in \{0, 1\}$, where $PR = 1(PR_1)$ represents the estimated affected region and $PR = 0(PR_0)$ represents the remaining regions. The binary map by ground truth (manual segmentation) represented by $GT \in \{0, 1\}$, where $GT = 1(GT_1)$ represents the actual affected region and $GT = 0$ (GT_0) represents the remaining regions. The brief details of the different metrics are as follows:

False Positive Volume Function (FPVF): It measures the misclassification during the segmentation, which is formulated as:

$$FPVF = \frac{|PR_1| - |PR_1 \wedge GT_1|}{|GT_1|} \tag{9}$$

where PR_1 represents the segmented outcome by proposed method and GT_1 denotes the ground truth. Total number of pixels in PR_1 and GT_1 is represented by $|PR_1|$ and $|GT_1|$, respectively.

False Negative Volume Function (FNVF): It measures the loss of significant elements in the segmented region, which is formulated as:

$$FNVF = \frac{|GT_1| - |PR_1 \wedge GT_1|}{|GT_1|} \tag{10}$$

Note that, lower value of FPVF and FNVF signifies promising results.

Dice Coefficient: Dice coefficient (DSC) is a statistic that measures the similarity through overlapping regions between ground truth and automated segmentation. It is formulated as:

$$SI = 2 \times \frac{|PR_1 \wedge GT_1|}{|PR_1| + |GT_1|} \tag{11}$$

Jaccard Index(JI): The pixel elements of two segmented outcomes are compared by the JI, which is formulated as:

$$JI = \frac{|PR_1 \wedge GT_1|}{|PR_1 \vee GT_1|} \tag{12}$$

If the similarity between two segmented outcomes is increased, then the JI value is also increased.

Sensitivity: The proportion of positives which are accurately recognized are estimated by sensitivity, which is represented as:

$$\text{Sensitivity} = \frac{|PR_1 \wedge GT_1|}{|GT_1|} \tag{13}$$

Specificity: The proportion of negatives which are accurately recognized are estimated by specificity, which is represented as:

$$\text{Specificity} = \frac{|PR_0 \wedge GT_0|}{|GT_0|} \tag{14}$$

5 Experiment Details

The performance evaluation of different segmentation methods [9, 13, 17, 18] have been compared with the proposed method based on the BraTS 2015 dataset [1, 32].

5.1 Data

BraTS 2015 dataset contains training and challenge datasets. The datasets are further categorized into high- and low-grade gliomas with ground truth. In BraTS 2015 training dataset, high-grade gliomas has 220 subjects, whereas low-grade gliomas has 54 subjects. Ground truth contains five segmentation classes, such as enhance tumor, non-enhance tumor, necrosis, edema, and normal tissues, whereas challenge datasets contain 53 subjects with unknown tumor grade. Ground truth is not provided in challenge datasets. Four modalities are available in each patient's brain, namely FLAIR, T1, T1c, and T2. All four MRI modalities are already co-registered and skull stripped. The brain slice thickest is 1 mm × 1 mm × 1 mm voxels. The example of four modalities MR images is shown in Fig. 6. Our training set contains 125 glioma subjects including high grade and low grade, which are further divided into training set with 100 glioma subjects and testing set with 25 glioma subjects. The dimension of each subject is 240 × 240 × 4 × 155. 5000 random patches with size 33 × 33 are chosen from each glioma subject. So, the total number of training patches is 5,00,000.

5.2 Parameters

Several parameters are involved in the present architecture, such as the size of the convolutional kernel, convolutional layer's feature maps number, pooling size, stride

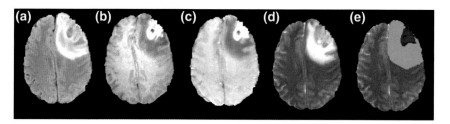

Fig. 6 Example of four different modalities of high grade brain tumor and ground truth from BraTS 2015 dataset. **a** FLAIR, **b** T1, **c** T1C, **d** T2, and **e** ground truth. Color codes: green - edema, red - enhanced tumor, blue - necrosis + non-enhanced tumor

size during convolution and pooling, and number of neurons in FC layer. These parameters have been tested on the proposed CNN architecture, which contains four convolutional layers with LReLU activation function, two max-pooling layers, and two FC layers with LReLU except for the last FC layer. The softmax activation function is applied to the outcome of the final FC layer. In all convolutional layers, we choose 3×3 kernels. We use 32 convolution filters for first two convolutional layers and 64 convolution filters for next two convolutional layers. Each of two max-pooling layers is occurring alternating order after each two convolutional layers. The number of neuron number in the last two FC layers are 128 and 5, respectively. In the last layer, number of neurons is same as the desired class number. Tables 1 and 2 show parameters value related to the proposed method. Besides the parameters in CNN architecture, image patch size also plays a very vital role during training. It is experimentally tested that for the patch size 33×33, the proposed architecture provides better segmented results.

5.3 Application to the Test Dataset

After optimizing the several CNN parameters during training, the present work applied the proposed architecture on 25 subjects, chosen from the BraTS 2015 test dataset. Figure 4 shows the architecture of the proposed CNN model. The results in Table 3 show the statistical measurement of the present work on BraTS 2015 dataset for three segmentation classes. Segmentation results of the proposed method on BraTs dataset are shown in Fig. 7. The mean dice scores values of the proposed method are 0.87, 0.79, and 0.77 for complete, core, and active tumor, respectively.

To measure the efficiency of the small size kernels over large size kernels, present work replaces two consecutive convolutional layers with 3×3 kernels by one convolutional layer with 5×5 kernels. So, in Table 1, layers 1 and 3 are replaced by one convolutional layer with 5×5 kernels with 32 feature maps, whereas layers 6 and 8 in Table 1 are replaced by one convolutional layer with 5×5 kernels with 64 feature maps. Using this new architecture (Table 4), although the number of layers

Table 1 Layer-wise parameters of the proposed architecture. In input and output column, first two dimensions denote the feature map/neuron size and last dimension refers the number of feature maps/neurons

L. no.	L. type	K. size	Stride	No. of Feas maps/FC units	Activation Func.	Input	Output
1	Convolution	3×3	1×1	32	–	$33 \times 33 \times 4$	$31 \times 31 \times 32$
2	Activation	–	–	–	Leaky ReLU	$31 \times 31 \times 32$	$31 \times 31 \times 32$
3	Convolution	3×3	1×1	32	–	$31 \times 31 \times 32$	$29 \times 29 \times 32$
4	Activation	–	–	–	Leaky ReLU	$29 \times 29 \times 32$	$29 \times 29 \times 32$
5	Max-pooling	2×2	2×2	–	–	$29 \times 29 \times 32$	$14 \times 14 \times 32$
6	Convolution	3×3	1×1	64	–	$14 \times 14 \times 32$	$12 \times 12 \times 64$
7	Activation	–	–	–	Leaky ReLU	$12 \times 12 \times 64$	$12 \times 12 \times 64$
8	Convolution	3×3	1×1	64	–	$12 \times 12 \times 64$	$10 \times 10 \times 64$
9	Activation	–	–	–	Leaky ReLU	$10 \times 10 \times 64$	$10 \times 10 \times 64$
10	Max-pooling	2×2	2×2	–	–	$10 \times 10 \times 64$	$5 \times 5 \times 64$
11	Fully connected (FC)	–	–	128	–	$1 \times 1 \times 1600$	$1 \times 1 \times 128$
12	Activation	–	–	–	Leaky ReLU	$1 \times 1 \times 128$	$1 \times 1 \times 128$
13	Fully connected (FC)	–	–	4	–	$1 \times 1 \times 128$	$1 \times 1 \times 4$
14	Activation	–	–	–	Softmax	$1 \times 1 \times 4$	$1 \times 1 \times 4$

Table 2 Few parameters of the proposed CNN architecture

Parameter	Value
Bias	random
Weight	random
Leaky ReLU	0.33
Dropout	0.5
Initial learning rate	0.01
Momentum factor	0.9
Decay factor	0.1
Batch size	500
No. of batches	1000
Epochs	35

is reduced, but the total number of weights is increased and performance is also reduced. Figure 8 shows the segmentation results by 3×3 kernels and 5×5 kernels along with ground truth. In Fig. 8, it is seen that 5×5 kernels based architecture miss-classifies few areas of edema and enhanced tumor regions, which is not actually present in the ground truth. Present work also measures the performance of the proposed method using ReLU activation instead of LReLU activation function. The segmentation performance of the present work using LReLU and ReLU activation functions with respect to ground truth is shown in Fig. 9.

Table 3 Few quantitative indices value for proposed method on BraTS 2015 training dataset

Quantitative indices	Regions		
	Complete	Core	Enhance (Active)
Dice similarity coefficient (DSC)	0.87	0.79	0.77
Jaccard Index	0.81	0.76	0.68
Sensitivity	0.86	0.78	0.71
Specificity	0.87	0.77	0.74
FPVF	0.04	0.09	0.10
FNVF	0.03	0.11	0.09

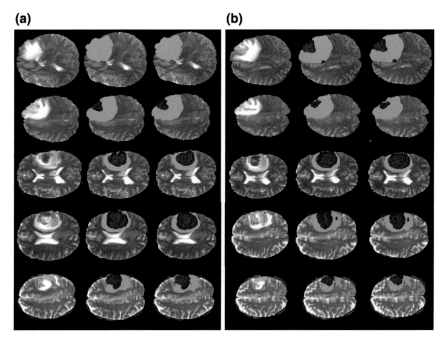

Fig. 7 Segmentation example on BraTS datsets. (**a, b**) T2 (first column), proposed segmentation (second column) and manual segmentation (third column). Color codes represent: red—enhanced tumor, green—edema, blue—non-enhanced tumor and necrosis

The evaluation results of the proposed method with respect to ground truth on BraTS dataset are shown in Fig. 7. The comparison results between different segmentation algorithms on BraTS dataset are shown in Table 5, which signifies the proposed CNN architecture achieves better performance than other methods [9, 13, 17, 18]. For implementing our proposed CNN architecture, MATLAB is used as a software tool.

Table 4 Layer wise parameters after replacing each two consecutive convolutional layers (with 3 × 3 kernels) of the proposed method (Table 1) by one convolutional layer (with 5 × 5 kernels). In input and output column, first two dimensions denote the feature map/neuron size and last dimension refers the number of feature maps/neurons

L. no.	L. type	K. size	Stride	No. of Feas maps/FC units	Activation Func.	Input	Output
1	Convolution	5 × 5	1 × 1	32	–	33 × 33 × 4	29 × 29 × 32
2	Activation	–	–	–	Leaky ReLU	29 × 29 × 32	29 × 29 ×32
3	Max-pooling	2 × 2	2 × 2	–	–	29 × 29 × 32	14 × 14 × 32
4	Convolution	5 × 5	1 × 1	64	–	14 × 14 × 32	10 × 10 × 64
5	Activation	–	–	–	Leaky ReLU	10 × 10 × 64	10 × 10 × 64
6	Max-pooling	2 × 2	2 × 2	–	–	10 × 10 ×64	5 × 5 × 64
7	Fully connected (FC)	–	–	128	–	1 × 1 × 1600	1 × 1 ×128
8	Activation	–	–	–	Leaky ReLU	1 × 1 × 128	1 × 1 × 128
9	Fully connected (FC)	–	–	4	–	1 × 1 × 128	1 × 1 × 4
10	Activation	–	–	–	Softmax	1 × 1 × 4	1 × 1 × 4

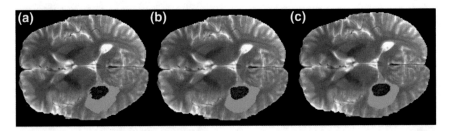

Fig. 8 Segmented results by 3 × 3 kernels (**b**) and 5 × 5 kernels (**c**) with respect to ground truth (**a**)

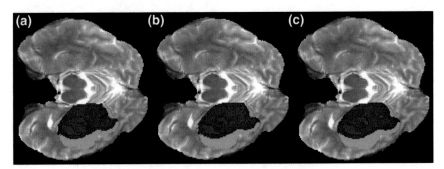

Fig. 9 Segmented results by LReLU activation (**b**) and ReLU activation (**c**) after 35 epochs with respect to ground truth (**a**)

Table 5 Comparison results on BraTS dataset for different segmentation techniques

Authors	Techniques	Accuracy (Dice Score)		
		Complete	Core	Enhance
Hussain et al. [18]	CNN	0.80	0.67	0.85
Dvorak et al. [9]	K-means and CNN	0.83	0.75	0.77
Zikic et al. [17]	CNN	0.83	0.73	0.69
Havaei et al. [13]	SVM	0.86	0.77	0.73
Proposed	Patch based CNN	0.87	0.79	0.77

6　Conclusion and Future Work

A convolutional neural network-based automated segmentation method is proposed in this present work for assisting in tumor detection of the brain. In the present work, we use small convolutional kernels (3×3 filters), LReLU as an activation function and dropout as a regularization method to overcome the problems of the traditional artificial neural network. The advantage of using 3×3 filters is to design a deep architecture that reduces the weights for each layer and more nonlinearity on the datasets by increasing the number of layers which makes a positive effect in overfitting. In this present work, the problem of ReLU is resolved by the LReLU activation function. The numerical and pictorial results of this present work signify that the proposed method achieved promising performance over other methods.

In future, the authors want to update the proposed model with more sophisticated CNN architecture to handle larger datasets through the GPU processing for higher performance with respect to time and accuracy in different tumor regions.

Acknowledgements This work is supported by the Board of Research in Nuclear Sciences (BRNS), DAE, Government of India under the Reference No. 34/14/13/2016-BRNS/34044.

References

1. Menze, B.H., Jakab, A., Bauer, S., Kalpathy-Cramer, J., Farahani, K., Kirby, J., Burren, Y., Porz, N., Slotboom, J., Wiest, R., et al.: The multimodal brain tumor image segmentation benchmark (BRATS). IEEE Trans. Med. Imaging **34**(10), 1993–2024 (2015)
2. Bauer, S., Wiest, R., Nolte, L.-P., Reyes, M.: A survey of MRI-based medical image analysis for brain tumor studies. Phys. Med. Biol. **58**(13), R97 (2013)
3. Bal, A., Banerjee, M., Sharma, P., Maitra, M.: Brain tumor segmentation on MR image using k-means and fuzzy-possibilistic clustering. In: 2018 2nd International Conference on Electronics, Materials Engineering & Nano-Technology (IEMENTech), pp. 1–8. IEEE, New York (2018)
4. Bal, A., Banerjee, M., Chakrabarti, A., Sharma, P.: MRI brain tumor segmentation and analysis using rough-fuzzy c-means and shape based properties. J. King Saud Univ.-Comput. Inf. Sci. (2018)
5. LeCun, Y., Bengio, Y., Hinton, G.: Deep learning. Nature **521**(7553), 436 (2015)

. 6. Srivastava, N., Hinton, G., Krizhevsky, A., Sutskever, I., Salakhutdinov, R.: Dropout: A simple way to prevent neural networks from overfitting. J. Mach. Learn. Res. **15**(1), 1929–1958 (2014)
7. Lyksborg, M., Puonti, O., Agn, M., Larsen, R.: An ensemble of 2D convolutional neural networks for tumor segmentation. In: Scandinavian Conference on Image Analysis, pp. 201–211. Springer, Berlin (2015)
8. Kleesiek, J., Biller, A., Urban, G., Kothe, U., Bendszus, M., Hamprecht, F.: Ilastik for multi-modal brain tumor segmentation. In: Proceedings MICCAI BraTS (Brain Tumor Segmentation Challenge), pp. 12–17 (2014)
9. Dvořák, P., Menze, B.: Local structure prediction with convolutional neural networks for multi-timodal brain tumor segmentation. In: International MICCAI Workshop on Medical Computer Vision, pp. 59–71. Springer, Berlin (2015)
10. Havaei, M., Davy, A., Warde-Farley, D., Biard, A., Courville, A., Bengio, Y., Pal, C., Jodoin, P.-M., Larochelle, H.: Brain tumor segmentation with deep neural networks. Med. Image Anal. **35**, 18–31 (2017)
11. Rajendran, A., Dhanasekaran, R.: Fuzzy clustering and deformable model for tumor segmentation on MRI brain image: a combined approach. Proc. Eng. **30**, 327–333 (2012)
12. Hamamci, A., Kucuk, N., Karaman, K., Engin, K., Unal, G.: Tumor-cut: segmentation of brain tumors on contrast enhanced MR images for radiosurgery applications. IEEE Trans. Med. Imaging **31**(3), 790–804 (2012)
13. Havaei, M., Larochelle, H., Poulin, P., Jodoin, P.-M.: Within-brain classification for brain tumor segmentation. Int. J. Comput. Assist. Radiol. Surg. **11**(5), 777–788 (2016)
14. Khotanlou, H., Colliot, O., Atif, J., Bloch, I.: 3D brain tumor segmentation in MRI using fuzzy classification, symmetry analysis and spatially constrained deformable models. Fuzzy Sets Syst. **160**(10), 1457–1473 (2009)
15. Selvakumar, J., Lakshmi, A., Arivoli, T.: Brain tumor segmentation and its area calculation in brain MR images using k-mean clustering and fuzzy c-mean algorithm. In: 2012 International Conference on Advances in Engineering, Science and Management (ICAESM), pp. 186–190. IEEE, New York (2012)
16. Rajendran, A., Dhanasekaran, R.: Brain tumor segmentation on MRI brain images with fuzzy clustering and GVF snake model. Int. J. Comput. Commun. Control **7**(3), 530–539 (2014)
17. Zikic, D., Ioannou, Y., Brown, M., Criminisi, A.: Segmentation of brain tumor tissues with convolutional neural networks. In: Proceedings MICCAI-BRATS, pp. 36–39 (2014)
18. Hussain, S., Anwar, S.M., Majid, M.: Brain tumor segmentation using cascaded deep convolutional neural network. In: 2017 39th Annual International Conference of the IEEE Engineering in Medicine and Biology Society (EMBC), pp. 1998–2001. IEEE, New York (2017)
19. Rao, V., Sarabi, M.S., Jaiswal, A.: Brain tumor segmentation with deep learning. In: MICCAI Multimodal Brain Tumor Segmentation Challenge (BraTS), pp. 56–59 (2015)
20. Wels, M., Carneiro, G., Aplas, A., Huber, M., Hornegger, J., Comaniciu, D.: A discriminative model-constrained graph cuts approach to fully automated pediatric brain tumor segmentation in 3-D MRI. Med. Image Comput. Comput.-Assist. Interv.-MICCAI **2008**, 67–75 (2008)
21. Pereira, S., Pinto, A., Alves, V., Silva, C.A.: Brain tumor segmentation using convolutional neural networks in MRI images. IEEE Trans. Med. Imaging **35**(5), 1240–1251 (2016)
22. Kaus, M.R., Warfield, S.K., Nabavi, A., Black, P.M., Jolesz, F.A., Kikinis, R.: Automated segmentation of MR images of brain tumors. Radiology **218**(2), 586–591 (2001)
23. Li, C., Gore, J.C., Davatzikos, C.: Multiplicative intrinsic component optimization (MICO) for MRI bias field estimation and tissue segmentation. Magn. Reson. Imaging **32**(7), 913–923 (2014)
24. Glorot, X., Bengio, Y.: Understanding the difficulty of training deep feedforward neural networks. In: Proceedings of the Thirteenth International Conference on Artificial Intelligence and Statistics, pp. 249–256 (2010)
25. Krizhevsky, A., Sutskever, I., Hinton, G.E.: Imagenet classification with deep convolutional neural networks. In: Advances in Neural Information Processing Systems, pp. 1097–1105 (2012)

26. Maas, A.L., Hannun, A.Y., Ng, A.Y.: Rectifier nonlinearities improve neural network acoustic models. In: Proceedings of the ICML, vol. 30, p. 3 (2013)
27. Hinton, G.E., Srivastava, N., Krizhevsky, A., Sutskever, I., Salakhutdinov, R.R.: Improving neural networks by preventing co-adaptation of feature detectors. arXiv:1207.0580 (2012)
28. Wong, S.C., Gatt, A., Stamatescu, V., McDonnell, M.D.: Understanding data augmentation for classification: when to warp? In: 2016 International Conference on Digital Image Computing: Techniques and Applications (DICTA), pp. 1–6. IEEE, New York (2016)
29. Fawzi, A., Samulowitz, H., Turaga, D., Frossard, P.: Adaptive data augmentation for image classification. In: 2016 IEEE International Conference on Image Processing (ICIP), pp. 3688–3692. IEEE, New York (2016)
30. Wang, J., Perez, L.: The effectiveness of data augmentation in image classification using deep learning. Technical report (2017)
31. Urban, G., Bendszus, M., Hamprecht, F., Kleesiek, J.: Multi-modal brain tumor segmentation using deep convolutional neural networks. In: MICCAI BraTS (Brain Tumor Segmentation Challenge. Proceedings, Winning Contribution, pp. 31–35 (2014)
32. Kistler, M., Bonaretti, S., Pfahrer, M., Niklaus, R., Büchler, P.: The virtual skeleton database: an open access repository for biomedical research and collaboration. J. Med. Internet Res. **15**(11) (2013)

Determining Perceptual Similarity Among Readers Based on Eyegaze Dynamics

Aniruddha Sinha, Sanjoy Kumar Saha and Anupam Basu

Abstract Understanding of a reading material depends on lexical processing and formation of perception. Differences between individuals exist due to various cognitive capabilities and psychological factors. In this paper, we focus on grouping of individuals according to their reading traits based on the eye movement behavior using affordable eye-tracking device. Detailed characterization of eye gaze, namely the distribution of the spatial spread and change in drift direction in fixations, is considered for analyzing the reading behavior. They relate to the perceptual span and attention, respectively. Analysis of variance (ANOVA) is performed with the features derived from the perceptual distance between pairwise individuals. Multidimensional scaling is used to obtain the relative perceptual orientation and attention among the individuals. Results demonstrate the feasibility of forming homogeneous groups based on the eye movement behavior.

Keywords Reading characteristics · Eyegaze fixation · ANOVA

A. Sinha (✉)
TCS Research & Innovation, Tata Consultancy Services, Ecospace 1B, Kolkata 700156, India
e-mail: aniruddha.s@tcs.com

S. Kumar Saha
Department of Computer Science & Engineering, Jadavpur University, Kolkata 700032, India
email: sks_ju@yahoo.co.in

A. Basu
Department of Computer Science & Engineering,
Indian Institute of Technology, Kharagpur 721302, India

A. Basu
National Institute of Technology, Durgapur 713209, India
e-mail: anupambas@gmail.com

© Springer Nature Singapore Pte Ltd. 2020
R. Chaki et al. (eds.), *Advanced Computing and Systems for Security*,
Advances in Intelligent Systems and Computing 996,
https://doi.org/10.1007/978-981-13-8969-6_7

1 Introduction

Reading and learning is a cyclic process of taking the visual inputs, processing the words and phrases, extracting the semantic information by linguistics processing, and finally building a perception of the content [1]. During this process, the understanding of the learning content not only depends on the amount of time spent in the reading but is affected by various cognitive and psychological factors [2]. These effects the eye movements, which are primarily controlled by cranial nerves. Though the control depends on the demand of mental process and readiness to move to the next part of the content [3], there are primarily two categories of belief, namely *global* and *direct* control. The theory of *global* control of eye movements states that the overall text difficulty of the content plays a dominating role rather than the local word-level properties [4]. On the contrary, the *direct* control states that the eyes look at only the word that is currently processed and do not move until it is completed [5].

The cognitive and psychological factors that affect during reading and under-standing of a content vary between the individuals. While performing a task, based on the difficulty level, the amount of working memory involved governs the mental workload and is termed as cognitive load [6]. There are various electrophysiological sensors [7] used for measuring the cognitive load. Among these, eye movement is a direct reflection of cognitive load, attention, and engagement in educational task [8] using an unobtrusive sensor.

With the recent popularity of distant learning [9], there is a growing need for personalization. In such scenario, the teaching and improvement plan demands for grouping of individuals with similar reading and learning skills. This not only would enable personalized training based on perceptual abilities but also would foster collaborative learning among like-minded individuals.

In this paper, we characterize a reader by observing the behavior of eye movements using low-cost and affordable eye tracker which would be more suitable for mass deployment. None of the above-mentioned work focuses on finding similarity between individuals based on the time-independent fixation characteristics. The novelty of this paper lies in the following

- Comparison between pair of individuals based on morphological structures of fixations and their distribution, namely spatial spread of a fixation, which gives individual perception during reading and number of change in drift direction within a fixation, which reflects attention.
- Relationship between individuals in terms of perceptual span and attention using multidimensional scaling (MDS) [10] in a higher-dimensional space.

The paper is organized as follows. Section 2 presents a brief review of the prior work. Section 3 gives the methodology of analyzing the fixation characteristics. The experimental setup, description of the textual content, participants details, and the data capture protocol are described in Sect. 4. Results are given in Sect. 5 followed by the conclusion in the final section.

2 Review of Prior Works

While doing a task, the experienced cognitive load is measured using various electrophysiological sensing [7] means, namely photoplethysmogram, galvanic skin response, electrocardiogram, skin temperature, electroencephalogram (EEG). Among these types of sensing, though EEG [11] provides direct information of neuronal responses in the brain, one needs to wear a device, whereas pupillary responses [12], namely fixation, saccades, and dilation, provide the information of activities in the cranial nerves using an eye tracker placed in front of an individual. While reading text material, fixations and saccades are done alternatively. The eyes remain fairly static for approximately 250 ms [13] during fixation. Fast ballistic movement between fixations is termed as saccades that lasts for 2–40 ms and spans over 7–9 characters [14]. Attention plays a major role in the understanding of textual content and drives the saccadic behavior [15]. Several models exist that characterize the behavior of eye movements. E-Z Reader model predicts saccadic behaviors in reading task [16]. Another major factor that affects the reading characteristics is mindless reading [17] where there is a decreased saccade duration, decreased fixation rate, and increased fixation duration [18]. With higher cognitive load during the reading of difficult texts, it is generally seen that the fixation duration and length of saccades increase [14].

Superior quality and costly infrared (IR) eye trackers are mostly used in content analysis. Eyegaze fixations are analyzed for evaluating the cognitive processes [19] during comprehension of a text using an eye tracker *EyeLink-1000* device. The effect of eye gaze while reading a second language is analyzed using *Tobii-X2-60* [20]. In the medical area, glaucoma patients are evaluated by eye tracking during a silent reading task [21]. Malnutrition affects the eye movement which is studied on the children using the *Tobii-X2-60* eye tracker [22]. However, for large-scale deployment in the scenario of e-learning, such costly eye trackers are not feasible to be used. During silent reading, for characterization of eye gaze, we find very few studies using low-cost eye trackers.

3 Methodology

Reading of a textual content involves understanding of the portion of the text currently focused on followed by seeking for new information by shifting the focus. Hence, fixations and saccades are the the primary eyegaze components during reading of a text. During fixation, the focus of the eye remains in a narrow region for 250–500 ms with the visual angle of less than $2°$ [23]. On the contrary, a saccade is characterized by a sudden shift in focus with visual angle of $210°–10°$ with high rotational velocity ($500°–900°/s$) [23]. Thus, the text processing during a reading task is enabled by alternate fixation and saccades. However, the nature of the fixation in terms of spatial spread, duration, and dynamics of the drift during a fixation varies

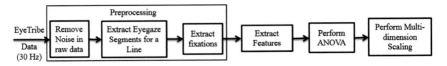

Fig. 1 Flowchart for deriving the perceptual relation between individuals during textual reading

between individuals. In the current work, we present a methodology to derive the perceptual distance between individuals based on their reading performance.

The block diagram of the overall process is shown in Fig. 1. Initially, as part of the preprocessing step, the noise cleaning of the eyegaze data is performed followed by detection of a line and detecting the fixations. Then, the features are extracted related to the fixation characteristics. ANOVA [24] is done between all the pair of individuals to find the difference in the perceptual space. Finally, a multidimensional scaling is performed to derive a relation between all the individuals in higher-dimensional space.

3.1 Preprocessing of the Eyegaze Data

Eyetracker device captures the eye gaze as $X - Y$ coordinates indicating the position of the display screen as focus of the eyes. EyeTribe [25], the eye tracker device used in the current experiment, is a low-cost infrared (IR)-based device. Hence, it is associated with large amount of inherent sensor noise. Moreover, the external IR interference and participants' head movements give rise to additional noise in the signal. Hence, initially the movement artifacts and sensor noise in the eyegaze data are removed using the preprocessing step.

Apart from the gaze location, the data captured from the EyeTribe provides the information on the signal quality using the state information-related metadata. Eye blinks and head movements lead to momentary loss of data in eye gaze which is interpolated using cubic spline [26]. Kalman filter and graph signal processing [25] are used to filter the noise in the eyegaze data.

After the noise cleaning, segments of the eyegaze data are extracted that corresponds to reading a particular line. As a line is read from left to right, the X-coordinate gradually increases and the Y-coordinate remains almost constant. While moving to the next line, as the eye gaze suddenly moves to the left side of the screen, there is a sudden decrease in X-coordinate. Thus as the text is read on the screen, the X-coordinate has an undulation pattern where the troughs are the starting of line. This is detected using a trough detection [27] algorithm.

Next, the fixations in each line are detected where the focus of the eyes remains momentarily static for approximately 250 ms [13]. A sliding window-based approach is utilized for the same where the change in mean value is used to detect the fixations [28]. The segment of the eye gaze corresponding to the fixation consists of a set of gaze locations containing spatiotemporal information.

3.2 Extraction of the Features

The gaze locations during a fixation and inter-fixation distance provide detailed information on the individual reading characteristics. Figure 2 demonstrate the eyegaze locations, scan path, fixations, and saccades. The speed of reading is primarily governed by the difficulty level of a content. In this paper, we are more interested in the individual reading trait which is inherent to an individual and independent of the type of content. Hence, two time-independent spatial features are derived from the eyegaze data, namely (i) spatial spread of fixation and (ii) number of change in direction within a fixation.

Spatial Spread of Fixation For a given fixation, a two-dimensional convex hull [29] is formed encompassing the gaze points within the time window of the fixation. A sample fixation in Fig. 2 is zoomed and shown the boundary represents the convex hull. Area within the convex hull is considered as a spatial spread feature, f_A^i of the ith of the fixation [30]. This size or area indicates the drift during a fixation indicating the perceptual span.

Direction Change Within a Fixation During a fixation, the eye gaze location drifts based on the stability of the neural control of the eye muscles. This is affected by the attention an individual has during a reading task. In Fig. 2, a sample fixation is zoomed and the scan path of the successive gaze points is shown. It can be seen that initially the gaze points move from left to right (shown in continuous line), then it moves from right to left (shown in dashed line), and finally, again it moves from left to right. Thus, the direction of the scan changes twice within the fixation. The number of change in direction is considered as a feature, f_n^i for the ith fixation.

Next for every individual, the feature set $f = \{f_A^i, f_n^i : 1 \leq i \leq N\}$ is derived for all the N fixations corresponding to the texts. These features reflect the perception and attention of an individual during the reading of a text. The NULL hypothesis is that the perception is similar for two individuals. Next for every feature, the ANOVA is done between the features of two individuals j and k. If the p value of the ANOVA is less than 0.05, then the NULL hypothesis is rejected indicating a significant separation between the two. In that case, the perceptual distance (S^{kl}) between the j and k is considered as F obtained from ANOVA. Otherwise, the perceptual distance (S^{kl}) is assumed to be a low constant value (C) indicating that j and k are quite close to

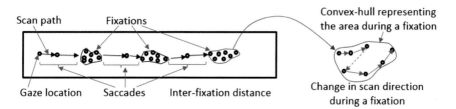

Fig. 2 Characteristics of fixation

each other. Thus, ($\frac{I(I-1)}{2}$) pairwise distance measures are computed among all the I individuals for each of the features.

$$S_A^{kl} = F, \text{ if } p < 0.05, \text{ where } [F, p] = \text{ANOVA}(f_A^j, f_A^k), \forall j \neq k$$
$$S_n^{kl} = F, \text{ if } p < 0.05, \text{ where } [F, p] = \text{ANOVA}(f_n^j, f_n^k), \forall j \neq k \qquad (1)$$
$$\text{else, } S_A^{kl} = S_n^{kl} = S_d^{kl} = C$$

3.3   Perceptual Relation Using Multidimensional Scaling

The pairwise distance measure is not good enough to provide relationship between all the individuals. This is due to the fact that even if distances between the individual pairs (i, j) and (j, k) are low the distance between (i, k) could be high. This is because of the relative positions in a some higher-dimensional space. In order to find the relative placement of all the individuals among themselves, the pairwise distance measures are converted to I points. This is done using MDS [31] such that the distances between the I points match with the experimental distances found from the eyegaze features. These I points are estimated in a high-dimensional space by least-square regression between the distance derived from the estimated points and the pairwise distances obtained from the eyegaze features. The error of this statistical fitment provides the goodness of fitment. Higher the number of dimension better is the fitment. In our experiment, we have found that the error is less than 0.005 (5% of minimum experimental distance) for two-dimensional space which is good enough for capturing the 95% of the variations [10]. This method provides a representation of relationship [10] between the individuals and help in finding the homogeneous groups.

4   Experimental Paradigm

Initially, we describe the experimental setup using the display and eye tracker along with the process of calibration. Then, the design of the reading content is presented. Finally, the participant details and the protocol of the experiment are given.

4.1   Setup

The participant sits at an approximate distance of 60 cm in front of 21 in. display screen, having a resolution of 1600 × 1200 pixels. The EyeTribe, connected to a computer using USB3 connector, is placed below the display facing toward the participant as shown in Fig. 3. Participants are asked to understand the text contents

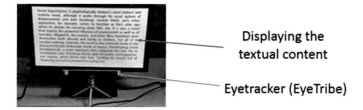

Displaying the
textual content

Eyetracker (EyeTribe)

Fig. 3 Experimental setup

which are shown on the display screen by reading in a silent manner. The computer captures the eye gaze and associated metadata from the EyeTribe at 30 Hz, in synchronous with the displayed texts. Throughout the experiment, the lighting condition is kept constant. In order to reliably capture the eyegaze data, participants are requested to minimize the extend of head movement.

Before the start of the reading task, the eye-tracking setup is calibrated using the software development kit (SDK) of EyeTribe. During calibration, participants are asked to follow a moving ball on the screen for approximately one minute. The quality of the calibration is computed by the SDK and indicated by a score within 1 and 5. It is desired to have a score of 5 (best quality) before the start of the experiment.

4.2 Description of the Stimulus for the Reading Task

In order to perform the reading task, textual contents are displayed on the display screen. The text is displayed using Calibri black font with a size of 32 points on a white background. Three paragraphs of textual contents are created using the texts based on novels,[1] demonetization, and nationalism.[2] The text is composed of approximately 132–164 words contained in 12–14 lines for each of the paragraphs. In order to capture the effect of the texts on the individuals, a focus is made on making the difficulty of the content on the higher side. This is because for an easy content, the reading behavior is usually similar, and as the difficulty levels increase, the behaviors emerge.

A benchmark on the difficulty level is performed using standard readability indices [32] namely Flesch–Kincaid Reading Ease (0–100), Flesch–Kincaid Grade Level (>3), SMOG Index (4–18), Coleman–Liau Index (>1). The range of these parameters is indicated in the parenthesis. Lower values of Flesch–Kincaid Reading Ease indicate greater difficulty level in the reading. On the contrary, the values of Flesch–Kincaid Grade Level, SMOG Index, and Coleman–Liau Index proportionately increase with the difficulty level of the text. For the designed texts, the mean values

---

[1] https://www.enotes.com/topics/great-expectations.
[2] https://www.quora.com/Why-is-there-so-much-criticism-for-demonetization-in-India-when-it-is-undoubtedly-a-correct-step.

for Flesch–Kincaid Reading Ease, Flesch–Kincaid Grade Level, SMOG Index are Coleman–Liau Index are 37.1, 14.6, 12.4, and 13.8, respectively [33]. Thus, the higher difficulty level is justified in the design of the text stimulus.

4.3 Participant Details

Fifteen engineering students (8 males, 7 females, age: 21–44 years) from an Engineering Institute, having similar cultural background, participated in the experiment. However, 6 of them had Bengali as first language and remaining as English during their high school. All the participants have normal vision or corrected to normal using glasses. Before using the data in the experiment, adequate anonymization is done. An informed consent form is also voluntarily signed by the participants following the Helsinki Human Research guidelines.[3]

4.4 Experiment Protocol

After the calibration of the EyeTribe setup, a fixation '+' is shown on the screen. Participants are requested to fixate their eye gaze on that '+' for 15 s. Next, the textual contents are displayed on the screen with one paragraph at a time. After a paragraph is read, a button is pressed by the participants to move to the next text. Before the start of the experiment, a demo session is given to the participants. During this time, they are given adequate time to get familiarized with the experimental procedure. At the end of the experiment, a 5-point Likert scale [34] is used to get feedback on the questionnaire related to concentration level while reading the text and whether or not they read English novels regularly.

5 Results

Among 15 participants, during their education, 6 had the first language as Bengali and remaining 9 had English. Hence, it is expected that the reading behavior would primarily have groups though there may be some variations among them. Moreover, during the feedback based on the questionnaire, participants indicated their concentration or attention level during the task. Seven participants mentioned that their concentration level was very high and remaining had moderate concentration level. The background and feedback are used to validate the findings from the eye movement behavior.

---

[3]https://www.helsinki.fi/en/research/research-environment/research-ethics.

The eyegaze data is analyzed for 15 individuals. Initially, for each individual, the fixation features for spatial spread (f_A) and direction change (f_n) are extracted for each of the three textual contents. Then, fixation features for each individual are combined for all the texts for further analysis.

Separate ANOVA for each of the features is done for pairs of individuals. The spatial spread reflects the perceptual similarity and direction change within fixation reflects the attention. There are 105 such pairs for 15 individuals. For the spatial spread feature f_A, the similarity S_A^{kl} measure is significantly different for 22 pairs. Thus, 74.3% of pairs have similar behavior for perceptual span. For the number of direction change feature f_n, the similarity S_n^{kl} measure is significantly different for 58 pairs, and for distance between fixation feature f_d, the similarity S_d^{kl} measure is significantly different for 47 pairs. Thus, 55.2% of pairs have similar attention. The distances between the pairs having significant difference are taken as the corresponding F values, and for the remaining, a very small constant value of $C = 0.1$ is considered.

Next, the distance information between the 105 pairs is used to derive the 15 points in a two-dimensional space using MDS. Figure 4 shows the relationship of perceptual span and attention between the individuals for both features f_A and f_n, respectively. The estimation error in MDS for spatial spread is 0.0009, and direction change count is 0.0022 indicating a fairly good estimation. This is because with a two-dimensional space, we are able to capture the 95% of the variations [10], where the error in distances derived from estimated points is within 5% (0.005) of minimum experimental distance (0.1).

In Fig. 4a, it is seen that for spatial spread f_A, 9 individuals are close to each other marked by a boundary. Out of these 9 individuals, 7 of them are having English as their first language. Thus, there is a 77.8% match between the background and the perceptual span. It can be seen that for remaining subjects with Bengali as first language, the behavior is distributed. This is possibly due to the fact that the change

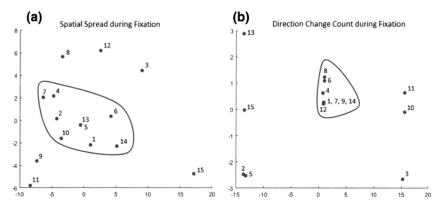

Fig. 4 Relationship between individuals for fixation features **a** spatial spread (f_A) indicating the perceptual span, **b** number of change in direction (f_n) indicating attention

in domain of primary language affects individuals differently as the other factors like memory, emotional state dominate.

For direction change count f_n, in Fig. 4b, it can be seen that 8 individuals are close to each other marked by a boundary. Out of these 8 individuals, 6 of them gave feedback that they were very much concentrated. Thus, there is a 75% match between the feedback and the attention level found from the eye gaze. For the remaining individuals, having moderate concentration, the points are distributed indicating that due to lack of attention, the eyegaze behavior is quite random.

Thus, the above results demonstrate the quantification of perceptual span from the distribution of the spatial spread in fixations and its relation with individuals' background. The distribution of the direction change count in fixations reflects the attention of an individual. These information can be used to create homogeneous groups and identify isolated individuals needing personalized focus. These groups can then be administered with similar learning improvement plans and possible scope for inter-group coordinations.

6 Conclusion

Eye movements during a reading task reflect the individual cognitive and psychological traits. The reading content is designed in such a way that the difficulty level is on the higher side so that it is useful to capture the difference in individual reading traits. Participants are chosen such that they have either Bengali or English as their primary language during high school. After the reading task, feedback is taken on whether they were able to fully concentrate during the task. Time-independent features, namely *fixation spread* and *change in direction count*, during a fixation are considered for the study. ANOVA is done on the features between pairwise individuals to capture the similarity in the eye movements. Multidimensional scaling on the pairwise similarity is performed to estimate the relative position among the individuals. Results indicate that individuals with English background are grouped based on *spatial spread* feature reflecting the perceptual span. On the contrary, the group formed by *change in direction count* feature reflects the attention level during the task. In the future, we aim to perform the study on a larger population and also extend the experiment on specific domains, namely physics, chemistry, mathematics.

Acknowledgements Authors would like to thank the participants for their cooperation during the experiment and data collection for the same.

References

1. McConkie, G.W., Rayner, K.: The span of the effective stimulus during a fixation in reading. Percept. Psychophys. **17**(6), 578–586 (1975)
2. Sweller, J.: Cognitive load during problem solving: effects on learning. Cogn. Sci. **12**(2), 257–285 (1988)

3. Carpenter, R.H.: Eye Movements, vol. 8 of Vision and Visual Dysfunction (1991)
4. O'regan, J., Lévy-Schoen, A., Pynte, J., Brugaillère, B.: Convenient fixation location within isolated words of different length and structure. J. Exp. Psychol.: Hum. Percept. Perform. **10**(2), 250 (1984)
5. Just, M.A., Carpenter, P.A.: The Psychology of Reading and Language Comprehension. Allyn & Bacon, Boston (1987)
6. Song, H.S., Pusic, M., Nick, M.W., Sarpel, U., Plass, J.L., Kalet, A.L.: The cognitive impact of interactive design features for learning complex materials in medical education. Comput. Educ. **71**, 198–205 (2014)
7. Haapalainen, E., Kim, S., Forlizzi, J.F., Dey, A.K.: Psycho-physiological measures for assessing cognitive load. In: Proceedings of the 12th ACM International Conference on Ubiquitous Computing, pp. 301–310. ACM, New York (2010)
8. Paas, F., Tuovinen, J.E., Tabbers, H., Van Gerven, P.W.: Cognitive load measurement as a means to advance cognitive load theory. Educ. Psychol. **38**(1), 63–71 (2003)
9. Mehlenbacher, B., Bennett, L., Bird, T., Ivey, M., Lucas, J., Morton, J., Whitman, L.: Usable e-learning: a conceptual model for evaluation and design. In: Proceedings of HCI International, vol. 2005, Citeseer (2005) 11th
10. Carroll, J.D., Chang, J.J.: Analysis of individual differences in multidimensional scaling via an n-way generalization of Eckart-Young decomposition. Psychometrika **35**(3), 283–319 (1970)
11. Khurana, V., Kumar, P., Saini, R., Roy, P.P.: Eeg based word familiarity using features and frequency bands combination. Cogn. Syst. Res. **49**, 33–48 (2018)
12. Just, M.A., Carpenter, P.A.: A theory of reading: from eye fixations to comprehension. Psychol. Rev. **87**(4), 329 (1980)
13. Hahn, M., Keller, F.: Modeling human reading with neural attention. arXiv preprint arXiv:1608.05604 (2016)
14. Rayner, K.: Eye movements in reading and information processing: 20 years of research. Psychol. Bull. **124**(3), 372 (1998)
15. Deubel, H., ORegan, K., Radach, R., et al.: Attention, information processing and eye movement control. Reading as a perceptual process, pp. 355–374 (2000)
16. Reichle, E.D., Pollatsek, A., Fisher, D.L., Rayner, K.: Toward a model of eye movement control in reading. Psychol. Rev. **105**(1), 125 (1998)
17. Feng, S., DMello, S., Graesser, A.C.: Mind wandering while reading easy and difficult texts. Psychon. Bull. Rev. **20**(3), 586–592 (2013)
18. Reichle, E.D., Reineberg, A.E., Schooler, J.W.: Eye movements during mindless reading. Psychol. Sci. **21**(9), 1300–1310 (2010)
19. Raney, G.E., Campbell, S.J., Bovee, J.C.: Using eye movements to evaluate the cognitive processes involved in text comprehension. J. Visualized Exp. JoVE (83) (2014)
20. Indrarathne, B., Kormos, J.: The role of working memory in processing l2 input: insights from eye-tracking. Bilingualism: Lang. Cogn. **21**(2), 355–374 (2018)
21. Murata, N., Miyamoto, D., Togano, T., Fukuchi, T.: Evaluating silent reading performance with an eye tracking system in patients with glaucoma. PloS one **12**(1), e0170230 (2017)
22. Forssman, L., Ashorn, P., Ashorn, U., Maleta, K., Matchado, A., Kortekangas, E., Leppänen, J.M.: Eye-tracking-based assessment of cognitive function in low-resource settings. Archives of Disease in Childhood (2016) archdischild–2016
23. Salvucci, D.D., Goldberg, J.H.: Identifying fixations and saccades in eye-tracking protocols. In: Proceedings of the 2000 Symposium on Eye Tracking Research & Applications, pp. 71–78. ACM, New York (2000)
24. Keselman, H., Huberty, C.J., Lix, L.M., Olejnik, S., Cribbie, R.A., Donahue, B., Kowalchuk, R.K., Lowman, L.L., Petoskey, M.D., Keselman, J.C., et al.: Statistical practices of educational researchers: an analysis of their anova, manova, and ancova analyses. Rev. Educ. Res. **68**(3), 350–386 (1998)
25. Gavas, R.D., Roy, S., Chatterjee, D., Tripathy, S.R., Chakravarty, K., Sinha, A.: Enhancing the usability of low-cost eye trackers for rehabilitation applications. PloS one **13**(6), e0196348 (2018)

26. De Boor, C., De Boor, C., Mathématicien, E.U., De Boor, C., De Boor, C.: A Practical Guide to Splines, vol. 27. Springer, New York (1978)
27. Jacobson, A.: Auto-threshold peak detection in physiological signals. In: Engineering in Medicine and Biology Society, 2001. Proceedings of the 23rd Annual International Conference of the IEEE, vol. 3, pp. 2194–2195. IEEE, New York (2001)
28. Veneri, G., Federighi, P., Rosini, F., Federico, A., Rufa, A.: Influences of data filtering on human-computer interaction by gaze-contingent display and eye-tracking applications. Comput. Hum. Behav. **26**(6), 1555–1563 (2010)
29. Graham, R.L., Yao, F.F.: Finding the convex hull of a simple polygon. J. Algorithms **4**(4), 324–331 (1983)
30. Rayner, K., Well, A.D., Pollatsek, A.: Asymmetry of the effective visual field in reading. Percept. Psychophys. **27**(6), 537–544 (1980)
31. Kruskal, J.B.: Multidimensional scaling by optimizing goodness of fit to a nonmetric hypothesis. Psychometrika **29**(1), 1–27 (1964)
32. Kincaid, J.P., Fishburne Jr., R.P., Rogers, R.L., Chissom, B.S.: Derivation of new readability formulas (automated readability index, fog count and flesch reading ease formula) for navy enlisted personnel. Technical report, Naval Technical Training Command Millington TN Research Branch (1975)
33. Sinha, A., Chaki, R., De Kumar, B., Saha, S.K.: Readability analysis of textual content using eye tracking. In: Advanced Computing and Systems for Security, pp. 73–88. Springer, Berlin (2019)
34. Hinkin, T.R.: A brief tutorial on the development of measures for use in survey questionnaires. Organ. Res. Methods **1**(1), 104–121 (1998)

High Performance Computing

2D Qubit Placement of Quantum Circuits Using LONGPATH

Mrityunjay Ghosh, Nivedita Dey, Debdeep Mitra and Amlan Chakrabarti

Abstract In order to achieve speedup over conventional classical computing for finding solution of computationally hard problems, quantum computing was introduced. Quantum algorithms can be simulated in a pseudo quantum environment, but implementation involves realization of quantum circuits through physical synthesis of quantum gates. This requires decomposition of complex quantum gates into a cascade of simple one-qubit and two-qubit gates. The methodological framework for physical synthesis imposes a constraint regarding placement of operands (qubits) and operators. If physical qubits can be placed on a grid, where each node of the grid represents a qubit, then quantum gates can only be operated on adjacent qubits, otherwise SWAP gates must be inserted to convert nonlinear nearest neighbour architecture to linear nearest neighbour architecture. Insertion of SWAP gates should be made optimal to reduce cumulative cost of physical implementation. A schedule layout generation is required for placement and routing a priori to actual implementation. In this paper, two algorithms are proposed to optimize the number of SWAP gates in any arbitrary quantum circuit. The first algorithm is intended to start with generation of an interaction graph followed by finding the longest path starting from the node with maximum degree. The second algorithm optimizes the number of SWAP gates between any pair of non-neighbouring qubits. Our proposed approach has a significant reduction in number of SWAP gates in 1D and 2D NTC architecture.

Keywords Quantum computing · Qubit placement · Quantum physical design

M. Ghosh (✉)
Department of Computer Science and Engineering, Amity University, Kolkata, India
e-mail: g.mrityunjay@gmail.com

N. Dey · D. Mitra
Department of Computer Science and Engineering, University of Calcutta, Kolkata, India

M. Ghosh · A. Chakrabarti
A. K. Choudhury School of Information Technology, University of Calcutta, Kolkata, India

© Springer Nature Singapore Pte Ltd. 2020
R. Chaki et al. (eds.), *Advanced Computing and Systems for Security*,
Advances in Intelligent Systems and Computing 996,
https://doi.org/10.1007/978-981-13-8969-6_8

1 Introduction

Quantum computing is a new computational paradigm to demonstrate the exponential speedup over classical non-polynomial time algorithms. Here probability and uncertainty replace determinism, in which energy can be delivered in discrete packets called quanta exhibiting dual nature to remain in particle form and also in wave form. Intrinsic features of quantum states like superposition and entanglement have made the system and its components fragile because whenever they interact with the environment, the information stored in the system decoherence thus resulting in error and consequent failure of computation [1]. To overcome the debilitating effects of decoherence and realize subtle interference phenomena in systems with many degrees of freedom, reliability should be enhanced by encoding a given computational state using blocks of quantum error-correcting code. Basic design principle of a fault-tolerant protocol is to avoid spreading out of a single qubit error due to fault gate or noise on a quiescent qubit to remaining qubits within one block of QECC [2, 3]. Fault-tolerant quantum error correction techniques include Shor fault-tolerant error correction, Steane error correction, Knill error correction, etc.

FT Quantum gates (single qubit or multiple qubit) are restrictive as they can only be applied on physically adjacent qubits. Various quantum techniques have been proposed for enabling various degrees of qubit interactions. 1D architectures are highly restrictive, since it can access only two neighbours per qubit, 2D architectures enable a qubit except for qubits present at the boundaries to access four adjacent neighbours and 3D architectures with six neighbours per qubit which has highly complex access method. Ion trap technology [4] uses 1-D interaction. Quantum dot (QD), superconducting (SC), neutral atom (NA) and photonics use 2-D interaction [5]. Cubic lattice crystal architecture in cellular automata uses 3-D interaction sequence of SWAP operations to couple any two non-adjacent distant qubits increases circuit latency and error rate [5, 6]. Amelioration of error threshold requires intricated control on quantum gate construction with higher fidelities followed by robust QECCs. In conventional VLSI design, the circuit placement starts with a weighted hypergraph where nodes represent standard cells and hyperedges represent connections among these cells. Circuit placement determines centre positions for nodes with a predefined size such that objective function-specific constraints can be optimized. Placement is followed by routing to connect placement cells through wires. Cost of computation in conventional VLSI technique relies upon wirelength, rate of power consumption and circuit delay. VLSI algorithms can be used to embed a weighted undirected interaction graph into a grid. But in qubit placement, positions of qubits keep varying at each iteration of SWAP gate insertion. Dynamic placement algorithms are to be devised so as to tackle time-variant nature of qubits to place it into a grid. By introducing dynamicity in placement, communication can be reduced [5]. After physically placing qubits in specific grid nodes using SWAP gates, exchange of qubits might be required in an ordered fashion to route two distant non-adjacent qubits towards each other in order to apply a quantum gate [5, 7]. Which, in turn, will have impact on all other qubits as their positions in placement grid will also be disturbed. Placement solution must be made reversible such that all moved qubits may return to their initial location by applying the same sequence of SWAP gates in reverse order [8]. Apart from considering mobility of qubits in placement grid in 2-D architecture, compatibility of n-qubit quantum gates should also be kept in mind, as they may not be directly implementable in a physical quantum machine. Consequently, gates must be further decomposed using a set of supported primitive quantum operators in the physical machine description (PMD) of the quantum machine. Realization of quantum gates on different physical quantum machines requires different number of primitive quantum operations. Physical realization of quantum computers is a function of unitary Hamiltonian operator to perform time-evolution of a closed quantum systems. Different quantum systems have different Hamiltonian, subsequently different PMDs. PMDs include quantum dot (QD) architecture where a qubit is represented by spin states of two electrons, superconducting (SC) where qubits are represented by charged carriers, ION trap (IT) where quantum system is based on a 2-D lattice of confined ions each representing a movable physical qubit within the lattice, neutral atom (NA) where trapped neutral atoms, those are isolated from the environment exploi00000.t quantum structure, linear photonics (LP) where a probabilistic two-photon gate is teleported into a quantum circuit with high probability and nonlinear photonics (NP) where quantum system is based on weak cross-Kerr nonlinearities, etc [9]. There exists a specific compatibility relationship between quantum gates and PMDs, viz. Controlled-NOT (CNOT), are supported in LP system, whereas SWAP gate is not available in NP physical machine description (PMD) [5, 9] (Fig. 1).

The complexity involved in the physical synthesis of quantum circuits can be effectively reduced by decomposing the overall synthesis into these stages [7]. Initial stage starts with a quantum algorithm containing both classical and quantum functions. Arbitrary quantum functions are difficult to synthesized, but they can be synthesized with the help of a quantum module Library QLib. Classical reversible functions can be synthesized using reversible logic synthesizer, whereas QLib is helpful to convert high-level quantum logic gates into low-level primitive quantum gates as it contains many commonly used quantum modules [10]. Quantum gate library can be optimized in terms of number of primitive quantum operations and associated delay by exploring one-qubit and two-qubit identity rules to remove redundancies in quantum gate implementation [5].

The circuit thus obtained after first stage is optimized in the next stage using an fault-tolerant quantum logic synthesis (FTQLS) which synthesizes and optimizes the non-FT logic to FT logic circuits. The last stage involves a direct synthesis of physical circuit by placing the qubits on a 2-D grid and routing cells properly to reduce communication overhead. In this stage, physical cost of implementing QECC is considered, as optimized cost will be a function of chosen QECC and corresponding PMD [5]. The remainder of this paper is as follows. Section 2 presents the earlier work related to physical design and synthesis of quantum circuits. Section 3 presents physical design-based problem statement as well as different approaches towards the problem. In Sect. 4 we have proposed our approach and a novel algorithm for optimizing SWAP gates in quantum circuits. Section 5 describes complexity analysis of our algorithm and last section will be the concluding part of this paper.

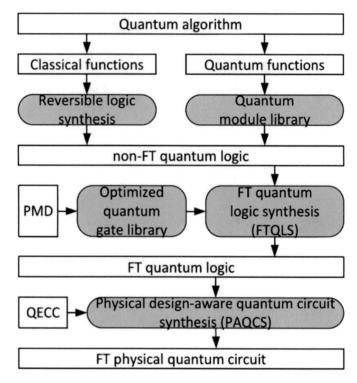

Fig. 1 Synthesis flow of quantum circuits[5]

2 Literature Review

Logical design phase of quantum circuits, in algorithmic level, assumes position-independent interaction of qubits. But, physical design phase, in implementation level, relies on neighbouring-qubit interactions [10]. Y. Hirata, M. Nakanishi and S. Yamashita first proposed an efficient method to convert an arbitrary quantum circuit to one on an LNN architecture applying permutation circuit for each qubit ordering [11].

Later, a trade-off between scalability and complexity is proposed by M. Saeedi, R. Wille and R. Drechsler incorporating nearest neighbour cost (NNC)-based decomposition methods [12]. A. Shafaei, M. Saeedi and M. Pedram formulated minimum linear arrangement (MINLA) using qubit reordering to improve circuit locality in an interaction graph [13]. Graph partitioning-based approach for LNN synthesis was proposed by A. Chakraborty et al. to provide significant reduction in the number of SWAP gates using reordering of qubit lines in quantum Boolean circuits (QBCs). Later, N. Mohammadzadeh et al. performed quantum physical synthesis applying netlist modifications through scheduled layout generation and iterative update scheduling using a gate-exchanging heuristic [14]. Later, scientists N. Mohammadzadeh and M. S. Zamani et al. proposed auxiliary qubit insertion technique after layout generation to meet design constraints using ion trap technology [15]. H. Goudarzi et al. presented a physical mapping tool for quantum circuits using trapped ion as PMD to generate the optimal universal logic block (ULB) which can perform any logical fault-tolerant (FT) quantum operation with minimum latency [16]. S. Choi and V. Meter first proposed an adder for 2D nearest neighbour with $\theta(\sqrt{n})$ depth and $O(n)$ number of qubits [17]. Based on the work presented by Choi and Van Meter, quantum addition on 2-dimensional nearest-neighbour architecture was proposed by M. Saeedi, A. Shafaei and M. Pedram where modified circuit structures for basic blocks of quantum adder were introduced to provide a significant reduction in communication overhead by adding concurrency to consecutive blocks and also by parallel execution of expensive Toffoli gates. The suggested optimizations, introducing consecutive block architecture, can improve depth from $140\sqrt{n} + K_1$ to $92\sqrt{n} + K_2$ for constant values of K_1 and K_2 [18]. Later, P. Pham and K. M. Svore presented a 2-dimensional nearest-neighbour quantum architecture for Shor's algorithm to factor an n-bit number in $O(\log^3 n)$ depth. Their proposed circuit incorporating algorithmic improvements (carry-save adders and parallelized phase estimation) and architectural improvements (irregular two-dimensional layouts and constant-depth communication with parallel modules)results in an exponential improvement in nearest-neighbour circuit depth at the cost of a polynomial increase in circuit size and width [19].

A. Shafaei, M. Saeedi and M. Pedram proposed optimization methods using mixed integer programming (MIP) formulation for time-variant dynamic placement of qubits. This approach places frequently interacting qubits as close as possible on the 2D grid to lessen the requirement of SWAP gates while routing [20]. Scalability to a large extent was achieved by Chia-Chun Lin and Susmita Sur-Koley in their work to design an effective physical design-aware quantum circuit synthesis methodology (PAQCS) incorporating quantum error-correcting code where two algorithms are proposed for qubit placement and channel routing, respectively. With the help of these two algorithms, the overhead of converting a logical to a physical circuit was reduced by 30.1%, on an average, relative to previous work [5].

3   Physical Design of Quantum Circuits

In order to simulate a quantum algorithm, physical realization of quantum circuits is required incorporating inherently reversible quantum gates. A reversible function establishes an one-to-one correspondence between input and output assignment where same number of variables are there in domain and range set. A circuit realizing a reversible function is a cascade of reversible quantum gates. Two common reversible gates include Controlled-Controlled NOT (Toffoli) gate and Fredkin gate. In multi-controlled Toffoli (MCT) from the domain of discourse, containing n-variables, $(n - 1)$ variables are treated as the control inputs and 1 variable is the target output which will be inverted iff all control lines are assigned to 1. If number of control inputs $(n - 1)$ is 2, then MCT is called Toffoli gate, and if $(n - 1)$ is 1, then MCT is called Controlled-NOT (CNOT) gate. Fredkin gate has n control lines and r target lines which will be interchanged iff the conjunction of all n input lines evaluates to 1. If a Fredkin gate does not have any control input, then it is called a SWAP gate. Quantum physical circuit architecture, invented so far, can process only single-qubit and two-qubit gates. Implementation of multi-controlled quantum gates into physical circuits is not feasible. So, decomposition of complex gates into a sequence of elementary quantum gates like NOT (Inverter), CNOT (Controlled-NOT), Controlled-V, Controlled-V^+(Inverse of Controlled-V), etc are required. Figure 2 shows decomposition of Toffoli and Fredkin gates into elementary CNOT, Controlled-V and Controlled-V^+ gates. During synthesis step involved in physical realization of Quantum circuits, optimization of circuit levels as well as gate count in a quantum Boolean network needs to be done [21]. Proper algorithmic approach is required for minimizing quantum cost of the circuit. Several performance metrics include:

- Number of lines and constant inputs: Initialization of quantum registers is complex because of the exponential state-space of an n-qubit register.
- Gate count and quantum cost: Number of elementary quantum operations needed to realize a gate contributes to quantum cost.
- Circuit depth: Number of steps required to execute all available gates in a circuit.
- Gate distribution: Coherence time for qubits and operation time for gates are widely affected by technological parameters as the total operation time of gates applied to a qubit must never exceed its qubit decoherence time. Otherwise the qubit value is lost before applying all gates [21].
- Nearest-neighbour cost: Most promising performance metric involved in physical realization of quantum circuit is the nearest-neighbour cost (NNC). In real quantum technologies, some restrictions exists between two interacting qubits. Most of the physical implementations follow linear nearest-neighbour (LNN) architecture where two qubits are allowed to interact if and only if they are adjacent to each other.

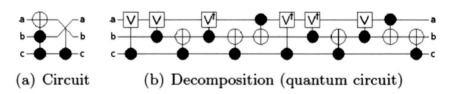

(a) Circuit   (b) Decomposition (quantum circuit)

Fig. 2 Decomposition of Toffoli and Fredkin gates

3.1   Linear Nearest-Neighbour (LNN) Synthesis

In order to minimize NNC, qubit lines must be reordered so that non-adjacent qubits can be adjacent before interaction. Without loss of generality, it is assumed that a given quantum Boolean circuit (QBC) is not in nearest neighbour form. In order to convert a given QBC to a corresponding LNN architecture, SWAP gates must be inserted appropriately. In Fig. 3, a Toffoli gate is shown which is not in LNN architecture. If qubit lines are not reordered, then number of SWAP gates in LNN representation is not optimal, whereas optimal solutions can be achieved after re-ordering of qubit registers resulting in less number of SWAP gates [22].

LNN synthesis for NOT, CNOT and Toffoli (NCT) and multi-controlled Toffoli (MCT) should be handled differently. The number of SWAP gates for a single qubit NOT gate is zero and for that of CNOT gate, it is simply the number of intermediate qubit lines between top and bottom control lines (Fig. 4).

Decomposition of multi-controlled Toffoli gates results in increased number of SWAP gates. For decomposition of a single C^k NOT gate, the number of TOFFOLI required is $2(k-1)+1$ and number of auxiliary qubits is $(k-2)$ [22] (Fig. 5).

The general idea of NNC optimization is to apply adjacent SWAP gates whenever a non-adjacent quantum gate occurs in the standard decomposition. SWAP gates are added in front of each gate with non-adjacent control and target lines to move a control line of a gate towards the target line until they become adjacent. In order to restore the original ordering of circuit lines, SWAP gates should be also added afterwards. A quantum gate 'g' where its control and target are placed at ath and bth lines, additional quantum cost of x. $| a - b - 1 |$ is needed. $| a - b - 1 |$ number of adjacent SWAP operations are required in order to make qubits adjacent, where x is the quantum cost for one SWAP operation [21, 22]. In order to minimize the number of SWAP gates, placement of qubits and establishment of a routing channel are two essential stages. In 2-dimensional nearest-neighbour, two-qubit gate, concurrent (NTC) architecture, any arbitrary circuit can be represented in a placement grid where maximum number of adjacent positions corresponding to a given cell is four. The goal of qubit placement is to place highly interactive qubits nearby in that grid and this

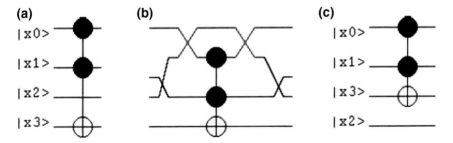

Fig. 3 Reordering of Toffoli gates for LNN synthesis **a** Toffoli gate in non-LNN architecture, **b** 4 SWAP gates without qubit reordering, **c** Reordering circuit with no requirement of SWAP gate

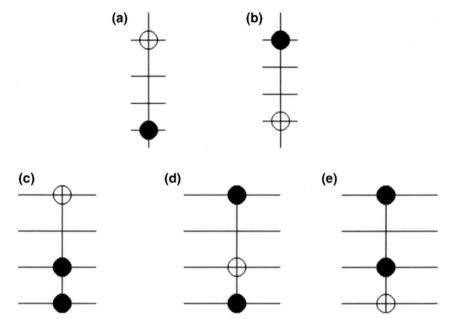

Fig. 4 Variation of MCT (**a**, **b**) Types of CNOT (NCT) (**c–e**) Types of TOFFOLI

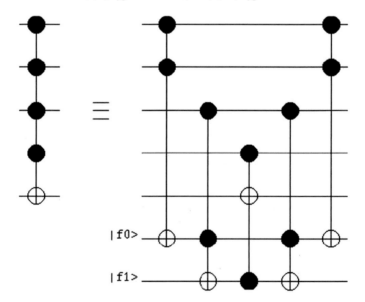

Fig. 5 Decomposition of MCT gate: replacement of a C^4NOT gate by equivalent Toffoli (NCTs) with two auxiliary qubits f_0, f_1

can be done with the help of an interaction graph, where the vertices refer to qubits and edge weights refer to the number of two-qubit gates operating on two-qubits. Consecutive gates in a given circuit can be executed in parallel due to sharing of control and target qubits which results in a working set of very few gates of a circuit at one scheduling level. This idea was incorporated by A. Shafaei et al. through providing a solution where the first phase starts with the formulation of m instances of the grid-embedding space on m sub circuits of an interaction graph for a circuit with N gates. The next phase is to insert SWAP gates before non-local two-qubit gates of each sub-circuit to obtain final placement. The final step requires a swapping network to align qubit arrangements of two consecutive sub circuits using 2-D bubble sort algorithm [20]. Another efficient qubit placement algorithm was proposed by Chia-Chun Lin et al. where their algorithm was based on breadth-first traversal [5]. The corresponding inputs to the logic circuits are the logic circuit and a parameter used as ranking of qubits to be chosen for their placement in neighbouring cells of a given qubit. In their work, they had taken degree of a vertex and activity of a vertex which is summation of all edge weights of its neighbouring qubits into consideration for determining priority of a qubit over other qubits. After selection of a vertex is over, that vertex is placed in the placement grid [5]. Now, say a vertex q_0 has more than four adjacent nodes q_1, q_2, q_3, q_4, q_5 and considering the grid to be vacant four high-priority nodes q_1, q_2, q_3, q_4 are placed on the grid at four neighbouring positions. There exists another node say q_6 which has adjacency with q_5 and rank of q_6 is k_6 and $k_1 > k_6 > k_{i,i=24}$. Since, node q_0 is deleted from priority queue so q_5 cannot be placed in adjacent positions in first iteration. So during the next iteration q_6 can never be placed though it has higher priority over adjacent nodes q_1, q_2, q_3, q_4. Thus, number of SWAP gates will increase as direct interaction between two adjacent nodes cannot be kept as that of logical circuit in the physical implementation. A different scenario can also happen during the physical implementation phase. The rank of a vertex (qubit) is determined by a function f which takes into account both the activity and degree of a node. Say, vertex q_0 is chosen first as its f is maximum. Now, consider a situation where a vertex q_2 is chosen with second maximum value where activity value is 100 and degree is 50, making $f = 150$ but there exists another vertex q_1 whose degree is 2 but activity value is 147 making $f = 149$. Since q_2 is chosen after q_0, so q_1 can never be chosen if q_1 is adjacent to only q_0 and there is no vacant position in the neighbouring cells of q_0 in the grid. In our work, we have proposed a technique to find the long path of highly interactive qubits where selection of vertices (qubits) will be made such that optimization in terms of selection of desired qubits can be made in order to provide reliability in the physical synthesis cost of any quantum circuit.

4 Proposed Approach

In this section, we are proposing two novel algorithms to optimize the number of SWAP gates based on edge weight optimization and removal of pair operations from the net-list, respectively. Given a quantum circuit Q, in its QASM form, we are looking for a optimal insertion of SWAP gates so that all gates g of Q can be executed in adjacent manner.

Result: A $n \times n$ grid, where all the qubits of a quantum circuit are placed by their index.

Input : QASM file Q, for a quantum circuit after FTQLS;

Initialization: set q =number of qubits in the quantum circuit;
set $G[i][j] = 0, \forall i, j = 1...q$, set $path[i] = 0, \forall i = 1...q$ and set $maxdeg = -1$;
set $n = \sqrt{q}$, and set $grid[i][j] = -1, \forall i, j = 1...n$;

Read Q **while** $!EOF$ **do**
 for each line L in Q;
 if *number of qubits in L $== 2$* **then**
  find index i and j of the two qubits $\in L$;
  set $G[i][j] = G[i][j] + 1$;
 else
  Discard L;
 end
end
set $i = 0, j = 0$ and $max = 0$;
while $i < q$ **do**
 while $j < q$ **do**
  if $G[i][j]! = 0$ **then**
   $path[i] = path[i] + 1$ and $path[j] = path[j] + 1$;
  else
   Continue;
  end
 end
end
set $j = 0$;
while $i < q$ **do**
 if $path[i] > j$ **then**
  set $j = path[i]$ and set $maxdeg = i$;
 else
  Continue;
 end
end
set $path[i] = -1, \forall i = 1...q$, set $path[0] = maxdeg$, and set $i = 1$;
while $i < q$ **do**
 set $j = -1$ and set *selected* =column index j of the element with maximum value among row $G[i - 1]$, where $j \notin path$;
 if $j! = -1$ **then**
  set $path[i] = selected$ and set $G[path[i - 1]][path[i]] = G[path[i]][path[i - 1]] = 0$;
 else
  set $path[i] = j$, where j is the row index of maximum value in G and $j \notin path$;
 end
 set $i = i + 1$;
end
set $path[0]$ at $grid[\lfloor (n - 1)/2 \rfloor][\lfloor (n - 1)/2 \rfloor]$ and the generated long path **spirally** to the $grid[n][n]$ matrix;

Algorithm 1: setLongPath() - generates interaction graph for a given quantum circuit and finds a long path to set the path in a grid

Result: QASM file Q, with optimally inserted SWAP gates.
Input : QASM file Q, for the quantum circuit after FTQLS and $grid[n][n]$;
Initialization: set q =number of qubits in the quantum circuit;
Read Q **while** $!EOF$ **do**

> for each line L in Q;
> **if** *number of qubits in L* $== 2$ **then**
>> find index i and j of the two qubits $\in L$;
>> find $x1$ and $y1$, where $grid[x1][y1] = i$;
>> find $x2$ and $y2$, where $grid[x2][y2] = j$;
>> Insert SWAP gate before L as follow (Considering $x1 <= x2$ and $y1 <= y2$): set $i = x1$;
>> **while** $i <= x2$ **do**
>>> INSERT SWAP($grid[i][y1]$, $grid[i+1][y1]$) before L;
>>> $i = i + 1$;
>>
>> **end**
>> set $i = y1$;
>> **while** $i < y2$ **do**
>>> INSERT SWAP($grid[x2][i]$, $grid[x2][i+1]$) before L;
>>> $i = i + 1$;
>>
>> **end**
>
> **else**
>> Discard L;
>
> **end**

end
Read Q **while** $!EOF$ **do**

> for each line $L1$ in Q;
> check the next line $L2$ in Q;
> **if** $L1$ *is identical to* $L2$ **then**
>> Goto the previous line $L0$ of $L1$;
>> Remove both the lines $L1$ and $L2$ from Q;
>> set $L1 = L0$;
>
> **else**
>> *Continue*;
>
> **end**

end

Algorithm 2: optimizeRoute() - optimize the number of SWAP gates during the SWAP operations

The First algorithm (**setLongPath()**) is generating an interaction graph from a given quantum circuit. The generated graph will denote the number of qubits of the given circuits as its set of vertices, the operations between different qubits as its set of edges and the number of binary operations between any two qubits as the weight of every individual edge.Then the algorithm takes the generated graph to find a long path starting with the node, having maximum degree. Then it sets the generated long path into a $n \times n$ grid. The starting node of the path, having maximum degree will be assigned to $(\lfloor (n-1)/2 \rfloor, \lfloor (n-1)/2 \rfloor)^{th}$ cell of the grid. If all the vertices are not present in the long path, then the algorithm takes remaining vertices to generate next long path in order to append it with the previous output. The placement of qubits in their respective cells of the grid will generate a long path in spiral fashion. Hence, the initial placement of qubits in a 2D architecture is done by this algorithm to optimize the number of SWAP gate insertions. The algorithm considers greedy approach to include new nodes into the long path. It checks the interactions between every pair of qubits in order to find the maximum interacting qubits among them. Our initial assumption is that the number of qubits is q. Hence, the complexity of the first algorithm is $O(q^2)$, where q is the number of qubits in the quantum circuit.

The second algorithm (**optimizeRoute()**) is designed to optimize the number of SWAP gates during routing step to achieve any quantum operation between two non-neighbouring qubits. It takes quantum assembly (QSAM) file with quantum gate list and the generated grid as input and inserts SWAP gates at the required position. Then it finds all the pair of consecutive gate operations, where the operation and operand(s) are same and removes that pair of operations from the QASM file. Hence, the gate cost is optimized. The algorithm reads each instruction from the generated QASM file and applies SWAP gates as needed. Therefore, the complexity of **optimizeRoute()** is $O(I)$, where I is the number of instructions in the generated QASM file by our first algorithm.

5   Proof of Correctness

Consider an interaction graph G consisting of sets of vertices (qubits), sets of edges (quantum operations) and respective edge weights (number of SWAP gates) is given below. Our placement algorithm setLongPath() chooses vertex, q_5 as the initial vertex as its degree of interaction is maximum (Fig. 6).

Then a greedy approach is applied to select the next vertex among the neighbouring vertices of the chosen vertex with the maximum edge weight as it represents maximum interaction in terms of requirement of SWAP gates.

The output of the algorithm will be an optimal path ensuring the reachability of all vertices (qubits) giving priority to highly interacting qubits. This will give us

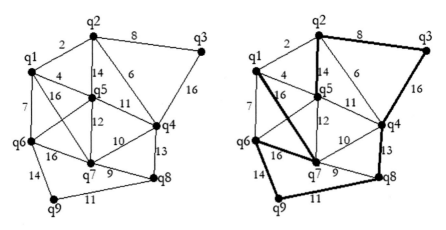

Fig. 6 Interaction graph of a quantum circuit and the long path on it

an optimal path $(q_5 - q_2 - q_3 - q_4 - q_8 - q_9 - q_6 - q_7 - q_1)$ which can be easily placed on a $n \times n$ grid where qubit placement will be stored from $(\lfloor (n-1)/2 \rfloor, \lfloor (n-1)/2 \rfloor)$th position of the grid and placement of remaining qubits will be done in spiral fashion giving us a linear path as shown in Figs. 7 and 8.

The initial grid configuration is represented in step $S1$. Steps $S2$, $S3$ and $S4$ are required to make non-adjacent qubits q_1, and q_9 adjacent to each other. Steps $S3'$, $S2'$ and $S1'$ are required to retain the initial configuration once operation between qubits q_1 and q_6 is performed. Now, if another quantum operation is defined between q_1 and q_6, then a communication channel is to be made up through steps $S1''$ to $S3''$. The final configuration is shown in steps $S4''$. But, this placement leads to non-optimal routing as it requires four additional SWAP gates which can be avoided through our algorithm optimizeRoute() as it removes unnecessary intermediate SWAP gates as show in Fig. 9. Thus, an optimized routing is obtained in physical synthesis of the given quantum circuit.

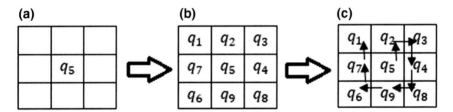

Fig. 7 Steps involved in placements of qubits in an $n \times n$ placement grid **a** Placement of initial vertex, **b** Placement of all vertices, and **c** Selection of Long path using setLongPath() algorithm. Hence the initial placement of qubits is done by our proposed first algorithm

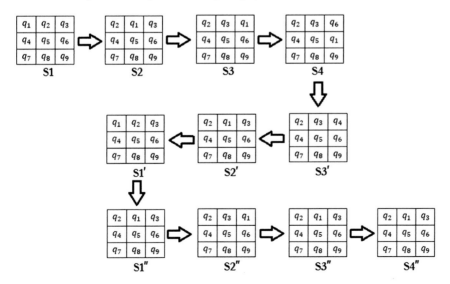

Fig. 8 Placement of qubits during routing in $n \times n$ grid

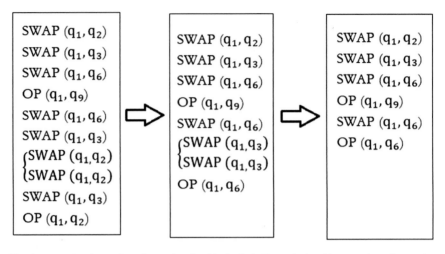

Fig. 9 Reduction in number of steps involved in OptimizeRoute() algorithm to reduce the number of SWAP gates to be inserted

6 Conclusion

In this work, we have discussed the issues regarding implementation of methodological framework of physical quantum circuit synthesis. The necessity of incorporating SWAP gates is also mentioned so as to make a non-LNN architecture behave as an LNN one. Our proposed algorithms setLongPath() and optimizeRoute() have led to significant reduction in the number of SWAP gates required and can be applied over any arbitrary quantum circuit. We will attempt to extend our research for achieving cost optimization in 3D NTC architecture.

References

1. Meier, F., Levy, J., Loss, D.: Quantum computing with spin cluster qubits. Phys. Rev. Lett. **90** (2003)
2. Hollenberg, L.C.L., Greentree, A.D., Fowler, A.G., Wellard, C.J.: Spin transport and quasi 2D architectures for donor-based quantum computing, Centre for Quantum Computer Technology School of Physics, University of Melbourne, VIC 3010, Australia
3. Maslov, D., Falconer, S.M., Mosca, M.: Quantum Circuit Placement. IEEE Trans. Comput.-Aided Des. Integr. Circuits Syst. **27**(4), 752–763 (2008)
4. Whitney, M., Isailovic, N., Patel, Y., Kubiatowicz, J.: Automated generation of layout and control for quantum circuits. In: Proceedings of the 4th International Conference on Computing Frontiers, pp. 83–94 (2007)
5. Lin, C.-C., Sur-Kolay, S., Jha, N.K.: PAQCS: physical design-aware fault-tolerant quantum circuit synthesis. IEEE Trans. Very Large Scale Integr. (VLSI) Syst. **23**(7) (2015)

6. Paler, A., Devitt, S.J., Nemoto, K., Polian, I.: Synthesis of topological quantum circuits. In: IEEE/ACM International Symposium on Nanoscale Architectures (NANOARCH), pp. 181–187 (2013)
7. DiVincenzo, D.P.: The Physical Implementation of Quantum Computation, IBM T.J. Watson Research Center, Yorktown Heights (2008)
8. Wille, R., Lye, A., Drechsler, R.: Optimal SWAP gate insertion for nearest neighbor quantum circuits. In: 2014 19th Asia and South Pacific Design Automation Conference (ASP-DAC) (2014)
9. Lin, C.-C., Chakrabarti, A., Jha, N.K.: Optimized quantum gate library for various physical machine descriptions. IEEE Trans. Very Large Scale Integr. (VLSI) Syst. **21**(11) (2013)
10. Maslov, D., Falconer, S.M., Mosca, M.: Quantum circuit placement: optimizing qubit-to-qubit interactions through mapping quantum circuits into a physical experiment. In: 44th ACM/IEEE Design Automation Conference (2007)
11. Hirata, Y., Nakanishi, M., Yamashita, S., Nakashima, Y.: An efficient method to convert arbitrary quantum circuits to ones on a linear nearest neighbor architecture. In: Proceedings of the 3rd International Conference on Quantum, Nano and Micro Technologies, pp. 26–33 (2009)
12. Saeedi, M., Wille, R., Drechsler, R.: Synthesis of quantum circuits for linear nearest neighbor architectures. Quantum Inform. Process. **10**(3), 355–377 (2011)
13. Shafaei, A., Saeedi, M., Pedram, M.: Optimization of quantum circuits for interaction distance in linear nearest neighbor architectures. In: Proceedings of the 50th Annual Design Automation Conference, pp. 41:1–41:6 (2013)
14. Mohammadzadeh, N., Sedighi, M., Zamani, M.S.: Quantum physical synthesis: improving physical design by netlist modifications. Micro- electron. J. **41**(4), 219–230 (2010)
15. Mohammadzadeh, N., Zamani, M.S., Sedighi, M.: Quantum circuit physical design methodology with emphasis on physical synthesis. Quantum Inform. Process. **13**(2), 445–65 (2014)
16. Goudarzi, H., Dousti, M., Shafaei, A., Pedram, M.: Design of a universal logic block for fault-tolerant realization of any logic operation in trapped-ion quantum circuits. Quantum Inform. Process. **13**(5), 1267–1299 (2014)
17. Choi, B.-S., Meter, R.V.: A $\theta(\sqrt{n})$-depth quantum adder on a 2D NTC quantum computer architecture. ACM J. Emerg. Technol. Comput. Syst. **8**(3), 24:–24:22 (2012)
18. Saeedi, M., Shafaei, A., Pedram, M.: Constant-factor optimization of quantum adders on 2D quantum architectures. In: Proceedings of the 5th International Conference on Reversible Computation, pp. 5–69 (2013)
19. Pham, P., Svore, K.M.: A 2D nearest-neighbor quantum architecture for factoring in polylogarithmic depth. Quantum Inform. Comput. **13**, 11–12 (2013)
20. Shafaei, A., Saeedi, M., Pedram, M.: Qubit placement to minimize communication overhead in 2D quantum architectures. In: Proceedings of the 19th Asia South Pacific Design Automation Conference, pp. 495–500 (2014)
21. Wille, R., Saeedi, M., Drechsler, R.: Synthesis of reversible functions beyond gate count and quantum cost. In: International Workshop on Logic Synthesis (IWLS) (2009)
22. Chakrabarti, A., Sur-Kolay, S., Chaudhury, A.: Linear nearest neighbor synthesis of reversible circuits by graph partitioning, CoRR, abs/1112.0564 (2011)

Debugging Errors in Microfluidic Executions

Pushpita Roy, Ansuman Banerjee and Bhargab B. Bhattacharya

Abstract The presence of a functionally correct reaction sequence graph has a significant advantage in the digital microfluidic (DMF) life cycle. Such a sequence graph is the basis from which the actuation sequence to be implemented on a target lab-on-a-chip is synthesized. In this paper, we investigate the possibility of using this sequence graph as a reference model for debugging erroneous reaction executions with respect to the desired output concentration. Our debugging method consists of program slicing with respect to the observable error in the golden implementation. During slicing, we also perform a step-by-step comparison between the slices of the erroneous output with other erroneous and error-free outputs. The reaction steps are then compared to accurately locate the root cause of a given error. Experimental results on the polymerase chain reaction (PCR) and linear dilution tree (LDT) protocols show that our method is able to pinpoint the errors.

1 Introduction

Microfluidic lab-on-a-chips (LoC) are set to replace cumbersome laboratory-based procedures and are being considered as the next-generation platform for on-chip implementation of biochemical laboratory assays [18]. The benefits of these devices have already been established in myriads of application domains, in DNA analysis, toxicity grading, molecular biology, and drug design, rapid and accurate

P. Roy (✉) · A. Banerjee
Indian Statistical Institute, Kolkata, India
e-mail: pushpita@isical.ac.in

A. Banerjee
e-mail: ansuman@isical.ac.in

P. Roy
Calcutta University, Kolkata, India

B. B. Bhattacharya
Indian Institute of Technology, Kharagpur, India
e-mail: bhargab.bhatta@gmail.com

© Springer Nature Singapore Pte Ltd. 2020
R. Chaki et al. (eds.), *Advanced Computing and Systems for Security*,
Advances in Intelligent Systems and Computing 996,
https://doi.org/10.1007/978-981-13-8969-6_9

diagnosis of various diseases including malaria, human immunodeficiency virus infection/acquired immunodeficiency syndrome, and for mitigating neglected tropical diseases prevalent in the developing countries [15].

Digital microfluidic (DMF) design automation tools take in a biochemical assay description expressed as a reaction sequence graph (Fig. 1) and translate it into a sequence of actuation operations for a target LoC architecture, which is loaded onto an on-chip controller, for further generation of actuation sequences for enabling various fluidic operations. During reaction execution on the LoC, embedded sensors placed below the electrodes [11] and/or overhead CCD cameras [14] are used to monitor reaction progress. Over the past decade, several automation methodologies and frameworks have been proposed to enable complete design flows from assay descriptions to LoC control [1, 5–8, 12, 13, 19–24].

With increase in the number, scale, and complexity of protocols being executed on a LoC, phenomenons like aging of electrodes, shorting of channels, and effects of such degradations are starting to surface. Due to significant re-use of the DMF grid, the electrical connections underneath the electrodes of the grid start to degrade. Some of the electrodes may not receive the electrical pulse to actuate the droplets, leading to blockages or slow actuations on the grid. These blockages may cause errors in assay execution at runtime when a droplet of the assay passes through the blocked region. Faulty input pins of mixers on the DMF LoC grid can lead to absence of input droplets at mixer input pins and produce an erroneous output or incorrect output volume after the mix operation. A mixer may become slower or faster and produce output later/earlier than the allotted time. Faults of the above kind are runtime manifestations. The actuation sequences synthesized from the golden implementation of the biochemical description for a target architecture are usually not fault-tolerant, and may thus lead to errors after completion of assay execution on the faulty LoC. The errors manifest as output volumetric errors at the end of execution or at intermediate detector locations. This needs to be debugged and fixed or isolated with respect to the desired output concentration factor specified as part of the golden reaction, before the next reaction is executed on the erroneous LoC.

In this paper, we study the problem of root-causing execution errors in protocol executions. In particular, we propose here an algorithmic method for error identification and error origin localization. Our approach takes in a golden implementation of a reaction sequence graph and error manifestations, and attempts to root cause the faulty operation and/or location on the LoC architecture. As output, our method produces a set of suspected locations and operations that may be possible origins of the fault. The foundation of our debugging approach is based on the concept of program slicing [2], a popular technique used in software engineering parlance for debugging program errors. Our proposed method includes two main steps. In the first step, we slice the assay description with respect to the error being debugged, to eliminate all those operations from the assay which do not affect the erroneous output directly or indirectly. This cuts down the suspected error region significantly since the assay may include many other operations which are not relevant to the fault manifestation, and thereby, do not help in the root-causing exercise. In the second stage, we make use of other erroneous manifestations and error-free outputs, if any.

We simultaneously compute the slices of these and compare them with the slice of the erroneous manifestation being debugged, to further localize the error source. A naive debugging method would involve debugging the entire assay execution on the LoC architecture, which may be an arduous time-consuming task for a reasonably complex protocol for a moderately sized LoC. Our method significantly cuts down the region of the assay to be examined.

Error-correcting methods for cyberphysical integration have been proposed in [14]. These methods aim to identify suspected erroneous regions online during an error occurrence in execution, and propose re-synthesis of the suspected error regions to include operations that can affect the detected fault. We believe that our method can complement such approaches and help in localizing the fault even better, and lead to lesser number of assay operations for re-synthesis. Defects in flow-based microfluidic chips have been studied in [9, 10], where a fault diagnosis technique using syndrome analysis is proposed. However, these are for flow-based chips. The objective of our paper is to explore the fault diagnosis problem in the DMF context. The problem of fluidic constraint violation checking that is addressed in [3] is not useful in our problem setting, since these methods work side by side with synthesis tools and do not consider execution errors in the LoC. The work in [16] discusses the problem of debugging security attacks using dynamic slicing. Authors in [25] used the notion of program slicing for inserting detection operations on a given input sequence graph and re-executing the subprogram which may cause the erroneous output. Our method not only identifies the set of operations of the given assay that may cause erroneous outputs but also isolates the particular operations that are erroneous.

To demonstrate the working of our method, we experimented on models of the linear dilution tree (LDT) [4] and the polymerase chain reaction (PCR) [17] protocols, with synthetically created random errors. In both the cases, we were able to localize the error source, as elaborated in Sect. 4.

This paper is organized as follows. Section 2 presents the problem statement, while Sect. 3 presents the methodology behind pruning and localization. Section 4 presents details of our implementation and experiments, while Sect. 5 concludes the discussion.

2   Problem Statement

Formally, we have (a) golden input sequence graph of a biochemical protocol, (b) the actuation sequence synthesized for a target DMF, (c) a set of erroneous outputs (at least one), and (d) a set of error-free outputs (if present). Our objective is to find the possible root-cause locations of the error in the assay. For the sake of simplicity, we consider a single fault (permanent or transient) in this work. In other words, the origin of the fault is a single location or operation or resource, while the manifestations may be errors in multiple outputs.

Consider the input bioassay P as shown in Fig. 1. We briefly explain the reaction graph. Two reagents (as depicted by nodes R_1 and R_2) are received as inputs to a mix

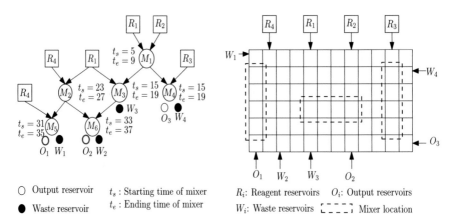

Fig. 1 Input sequence graph of an input bioassay and a target grid

operation M_1, where they are mixed for 4 units of times, starting from time unit 5, ending at 9. It may be assumed that time units 1–4 are spent in dispensing and moving the dispensed droplets to the input ports of the mixer. One of the outputs produced after the mix–split operation M_1 is then sent as input to the mix operation M_4, along with a newly dispensed droplet, shown in the figure as R_3. The mix operation M_4 continues from time unit 15 to 19, thereby producing 2 output droplets at the end of the mix–split operation. The rest of the operations are similar in nature and have similar interpretations. R_is represent the dispensing of reagents, M_is the mix operations, and O_is the outputs. Consider a biochip of dimension 6×12 on which the protocol is executed. Reagent reservoirs are attached with the cells $(1, 2)$, $(1, 5)$, $(1, 8)$, and $(1, 11)$, the waste reservoirs are with the cells $(3, 1)$, $(6, 3)$, $(6, 5)$, and $(3, 12)$, and the output reservoirs are with the cells $(6, 1)$, $(6, 8)$, and $(6, 12)$.

After execution of the input assay, let us suppose an erroneous output (incorrect concentration/volumetric ratio) is detected at the output reservoir $(6, 8)$. The erroneous outputs O_1 and O_2 are shown in Fig. 1 as red circles, while the other output O_3 is error-free. We attempt to find the root cause of the error manifestation in O_2 at $(6, 8)$.

3   Debug Methodology

Our debugging method consists of three main steps, as described below.

- *Time frame expansion*: This creates a time-annotated reaction graph based on the given reaction graph and the actuation sequence on the target LoC.
- *Slice computation*: This helps in pruning a given reaction graph to discard operations irrelevant to a given operation.
- *Error localization*: This involves comparison between the slices for more accurate localization.

Time frame expansion: The objective of this step is to instantiate the given reaction graph (e.g., in Fig. 1) on the biochip architecture, and unfold the actuation sequence over time, thereby creating a graph with time-annotated operations, as in Fig. 1. Figure 2 shows the transformed input sequence graph, where the square boxes represent the set of dispense operations and circular nodes represent the set of operational vertices of the assay. An edge from one vertex to another represents a dependency between them. Some nodes are annotated with a time interval which depicts the starting and ending time of the operation. Each reagent instance in Fig. 1 is replaced with a corresponding dispense operation with a unique identifier ($D_1, D_2, \ldots, D_4$). Each mix operation is also uniquely identified by an identifier ($M_1, M_2, \ldots, M_6$). The figure shows another type of circular node labeled as mv, which represents a move operation that depicts the movement of a droplet from a location (x, y) to another location (x', y') on the DMF grid. Each operation in the input sequence graph is annotated with the start and end times. The time frame expansion of the original reaction graph gives us a better artifact to debug, and we use it in our method. The output erroneous droplets are shown in red in Fig. 2.

Slice computation: The objective of the slicing operation is to prune out all the operations that have no influence on the erroneous output. The notion of a slice [2] has been extensively used in the software engineering community. We adopt the concept of the slice in the DMF context and show how it serves as an effective debugging aid. Given a time frame-expanded graph and an error location, we compute the backward

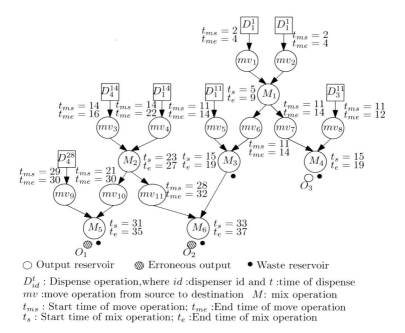

○ Output reservoir   ◍ Erroneous output   ● Waste reservoir

D_{id}^t : Dispense operation, where id :dispenser id and t :time of dispense
mv :move operation from source to destination M: mix operation
t_{ms} : Start time of move operation; t_{me} :End time of move operation
t_s : Start time of mix operation; t_e :End time of mix operation

Fig. 2 Unfolded input sequence graph of an input bioassay

slice to collect all nodes (i.e., operations in the graph) that have a direct or transitive influence on a given node. The backward slice computation attempts to compute the transitive closure of the dependencies encountered in the path from the slicing point to the start of the assay. The first step in slice computation is to identify a slicing criterion. We first consider the error manifestation output as the slicing criteria and compute its backward slice as below. The inputs to the slicing step are:

Algorithm 1: sliceComputation(c)

Unmark all operations in the assay and initialize slice to ϕ
for *all operations OP in the given assay executed before slicing point c* **do**
    if $(OP = dispense)$ **then**
        destLocation = dispense($time$)
        if $(destLocation = c)$ **then**
            Add operation OP in the stack S ;
            Return;
        end
    end
    if $(OP = move)$ **then**
        destLocation = move($src, time$)
        if $(destLocation = c)$ **then**
            Add operation OP in the stack S
            Call sliceComputation(src)
        end
    end
    if $(OP = mix)$ **then**
        destLocation = mix($input1, input2, time$)
        if $(destLocation = c)$ **then**
            Add operation OP in the stack S
            Call sliceComputation($input1$)
            Call sliceComputation($input2$)
        end
    end
end
Print all marked statements in the assay as the slice

– A time frame-expanded input sequence graph of an assay P
– A slicing criteria, in this case an erroneous output.

This step outputs a fragment of the assay P that is likely to be the source for the observed error at the output reservoir. Algorithm 1 presents the algorithm for computing slice for a given assay P. We explain its working philosophy. The algorithm sliceComputation takes one argument c as input, where c is the slicing criteria, which is initialized by the location of the erroneous output. The algorithm traverses the time frame-expanded graph and checks all operations of the assay that could be executed before the slicing point c and pushes those operations on a stack S that stores all operations that can affect the slicing criteria directly or indirectly. The operations on the assay are dispense, move, and mix. The *dispense* (*time*) operation dispenses a droplet on a cell at a certain time and produces a droplet at the dispense location as output. The *move*(*src*, *time*) operation takes two arguments: *src* refers to the source location, and *time* refers to the time required for the operation and produces a droplet at the *destLocation* as output. The *mix*(*input*1, *input*2, *time*) operation takes three arguments, where *input*1 and *input*2 refer to the location of the two input droplets and *time* refers to the time instance of the operation. It produces two output droplets as a result of the mix–split operation. For each of these droplets, the mix operation constitutes the slice. For a mix operation, the two input port locations constitute the slice. For each such input port, the corresponding move operation (which produces a droplet at that location) or a dispense (which dispenses a droplet at that port location) constitutes the slice. Similarly, for a move operation, considering the *destLocation* as the slicing criterion, the source constitutes the slice. If *destLocation* matches with the slicing criteria c, then the operation is added to the slice and the algorithm slice Computation is recursively called for the input location of that added operation. The slice computation is invoked starting from the slicing criterion c. The objective is to mark all nodes in the assay which affect c directly or transitively. To this effect, c is marked, and the slice of c is computed first. Nodes in the slice of c are marked and pushed onto a stack, since these need to be examined for slice computation now. This is done by recursively invoking the slice computation on them, as earlier by popping them off the stack in order and then marking and pushing their slices. The recursion terminates when a dispense method is encountered. In the process, the Algorithm 1 is able to collect all operations in the assay that affect the slicing criterion. It is easy to see that nodes in the graph which do not have a direct or transitive influence on c will *not be marked* in any step and thus we have a much smaller fragment of the assay to look at, for possible error localization. The slices for outputs O_1, O_2, and O_3 are shown in Fig. 3a–c. An illustration of the resulting slice follows below. We explain the slice building process and assay pruning operation with respect to O_2. Figure 4 shows the slice for O_2. The top-level node showing a *volumetric error at output O_2* is the root of the slice. The volumetric error at O_2 occurs either due to *the incorrect detection at output reservoir* or due to *the unsuccessful completion of the mixer M_6*. The event *unsuccessful completion of mixer M_6* occurs either due to an *Error at M_6* or due to the *Absence of the input droplets at M_6*. The event *Absence of the input droplets at M_6* occurs either for the event *Absence of the input droplet 1 at M_6* or for the event *Absence of the input droplet 2 at M_6*. The event *Absence of the input droplet 1 at M_6* occurs either for the event *Error at move operation* or for the event *Unsuccessful completion of mixer M_2*. Another event *Absence of the input droplet 2 at M_6* occurs either for the event *Unsuccessful completion of mixer M_3* or for the event *Error at the move operation*. Continuing similarly, we finally get the slice for O_2 in Fig. 4. ∎

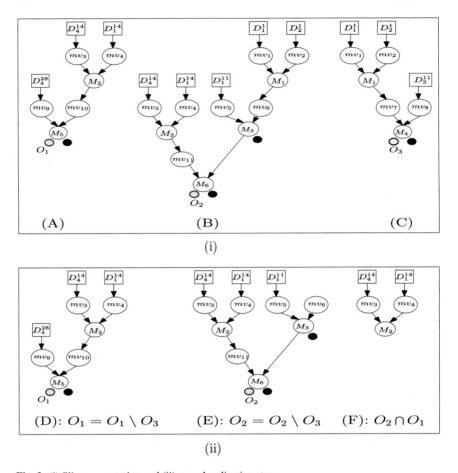

Fig. 3 (i) Slice computation and (ii) error localization steps

Error localization: Once the slice for the erroneous output is computed, we have a subset of assay operations affecting the error, and any operation in this subset can be a potential source of the error. Once the slice computation is over, the designer can be provided the marked statements in the resulting slice to examine the root cause of the error. Evidently, there may still be quite a few statements in the slice that the designer needs to look at. We now propose a couple of further optimization steps that can cut down the region even lesser. For this, we assume a *single-fault model*, which has been a popular fault model in the electronic design automation industry. We show that it leads to more efficient localization. We build the slices for all assay outputs. This may include other erroneous outputs and error-free outputs. The main motivation is as follows.

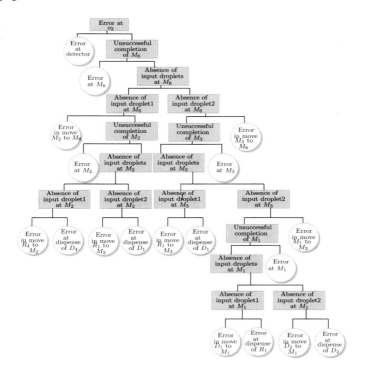

Fig. 4 Slicing in action for o_2

- An operation common between an erroneous output and an error-free output cannot be a potential source, since it would have otherwise affected the error-free output as well. Thus, all such operations can be safely *discarded* from the slice of the erroneous output to have a smaller suspected error region.
- Since we have a single source of error, operations common between the slices of erroneous outputs are potential candidates as the source. We can further prune down the slice of the error being debugged by *including* only those operations which are common with other erroneous outputs. This cannot be empty since we have a single-fault and multiple erroneous manifestations.

For both the cases above, the optimization depends on the availability of appropriate outputs (erroneous or error-free).

In case we have multiple manifestations being debugged right from the start, we compute the slices of each and then carry out the following steps:

- Difference computation between erroneous and error-free outputs.
- Intersection between slices of erroneous outputs.

Algorithm 2 describes the procedure for error localization that accepts two parameters, the number of outputs (m) and the number of erroneous outputs (n) of the assay. The set of outputs of the assay is divided into two groups. One is the set of erroneous outputs $\{O_1, O_2, \ldots, O_m\}$, and the other is the set of error-free outputs $\{O'_{m+1}, O'_{m+2}, \ldots, O'_n\}$. We build the slices for each erroneous output and for each error-free output. Generally, the slice for each erroneous output contains all possible candidate error sources. We then update the slice of each erroneous output by discarding the common operations that are present in the slices of both the erroneous outputs and the error-free outputs, as shown in steps 8 to 10 of Algorithm 1. Again, in steps 11 to 13, we find the intersection of all the updated erroneous output sets for a further optimization. We finally print the resulting slice E as the probable error sources. Figure 3a–c show the slices for outputs O_1, O_2, and O_3, respectively, in which, O_1 and O_2 are the erroneous outputs and O_3 is the error-free one. We want to debug the erroneous output O_2. Figure 3d shows the resultant slice of O_2 after the difference computation with the error-free output O_3. There are two erroneous outputs in the given assay. We compute the final slice as the non-empty intersection of the resultant slice of output O_2 with the other erroneous output O_1 to produce the probable error location. Figure 3e shows the probable error location and output of our method ErrorLocalization and the final slice with five nodes.

Algorithm 2: ErrorLocalization(n, m)

Erroneous set consists of $\{O_1, O_2, \ldots, O_m\}$ and error-free set consists of $\{O'_{m+1}, O'_{m+2}, \ldots, O'_n\}$;

for *all erroneous outputs* $O_1, O_2 \ldots O_m$ **do**
  | build slices $T_{e_1}, T_{e_2}, \ldots T_{e_m}$ respectively;
end

for *all error-free outputs* $O'_{(m+1)}, O'_{(m+2)} \ldots O'_n$ **do**
  | build slices $T'_{e_{(m+1)}}, T'_{e_{(m+2)}}, \ldots T'_{e_n}$ respectively;
end

for $i = 1$ *to* m **do**
  | **for** $j = 1$ *to* $(n - m)$ **do**
  | | $T_{e_i} = T_{e_i} \setminus T'_{e_j}$;
  | **end**
end

for $i = 1$ *to* m **do**
  | $E = \bigcap T_{e_i}$;
end

Print E as the probable error locations;

---

4   Implementation and Results

Our implementation takes as input a golden assay, the actuation sequence, and the output error. It produces a set of operations that may be the error source. We first discuss our experiments on LDT. Figure 5 shows the LDT input graph.

Fig. 5 Input sequence graph for LDT

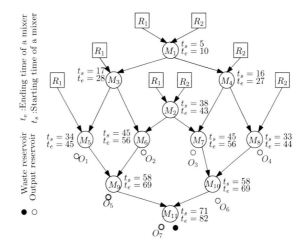

We considered a 8 × 13 grid as the target biochip architecture and an actuation sequence consisting of 19 operations produced by a synthesis tool. We inject errors at different locations of the assay, one at a time (single-fault model). We describe below one such experiment, where error was injected at mixer M_9. Effect of the faulty mixer is seen at outputs O_5 and O_7 which are erroneous and shown as red circles in Fig. 5.

We carried out our steps of slice computation followed by differencing and intersection. Figure 6 (i) and (ii) depicts the results of the different steps. Figure 6 (i) f shows the final slice, which points toward a fault in M_9. Our method is able to cut down the size of the faulty suspect region significantly in this case, finally reducing to a single node. This shows its effectiveness. We randomly varied the fault location to test the application of our method. Column 3 shows the number of nodes present in the original assay, while Column 5 shows the number of nodes present in the final slice.

A similar fault injection and debug experiment (as shown in Table 1) was carried out for the PCR streaming protocol. In this case as well, our method could reduce the faulty suspect region considerably.

Further, we varied the size of the grid for both LDT and PCR and the fault location simultaneously. We injected fault at different locations of the assay and analyzed the performance of our tool. In each case, the size of the faulty region was significantly less as compared to the original assay. Performance experiments with our tool are discussed in Table 2. The first column of the table shows the type of the assay. Column 2 and Column 3 show the index of the experiments of each protocol and the grid sizes of each protocol, respectively. Column 4 shows the injected error location for each experiment, and Column 5 shows the erroneous output locations. Finally, the last two columns of the table, respectively, show the time needed for the pruning method and the peak memory required for the respective experiments.

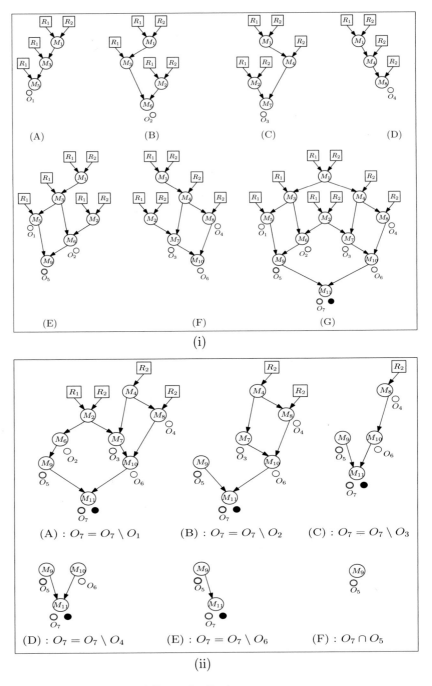

Fig. 6 (i) Slice computation and (ii) error localization steps

Table 1 Experimental results on LDT and PCR

Type of assay	Erroneous output	Nodes present on assay	Slice contains	After pruning remaining nodes
LDT	(8, 7)	19	19	1
	(8, 1)	19	7	4
	(8, 7)		19	
	(8, 11)		12	
	(8, 3)	19	12	3
	(8, 5)		9	
	(8, 7)		12	
	(8, 3)	19	12	8
	(8, 5)		9	
	(8, 7)		12	
	(8, 9)		9	
	(8, 11)		12	
PCR streaming	(8,3)	28	17	2
	(8, 8)	28	15	2
	(6, 3)	28	15	5
	(8, 3)		15	
	(8, 3)	28	15	4
	(8, 8)		15	

We now compare our results with the one in [14]. In this work, the authors have proposed an error recovery technique that re-synthesizes the suspected error region. For finding the suspected error region starting from an error manifestation, the slice of the error is computed. In Table 1, we have shown the number of statements that remain as suspected error origin locations, after the slice is computed. Our debugging method not only computes the slice of the erroneous output but also performs a slice comparison step for pruning the nodes which cannot affect the erroneous output. Hence, our method provides lesser number of nodes in the final result as possible origins of the erroneous output, as evident from the numbers in the final column of the same table.

5   Conclusion

In this paper, we have proposed an error debugging method that can efficiently localize errors in DMF executions. Our method is completely automatic and can localize the earliest error location in the erroneous assay using backward slicing and slice comparison. As future work, we wish to examine the applications and suitability of our method on more DMFB protocol realizations. While the method

Table 2 Performance records on LDT and PCR with varying grid sizes

Type of assay	Index	Grid size	Injected error	Erroneous output	Time for pruning (s)	Peak memory (MB)
LDT	1	8×13	M_{11}	(8, 7)	61	0.177
			M_8	(8, 1)	120.68	0.177
				(8, 7)		
				(8, 11)		
			M_6	(8, 3)	360.85	0.177
				(8, 5)		
				(8, 7)		
			M_2	(8, 3)	48	0.158
				(8, 5)		
				(8, 7)		
				(8, 9)		
				(8, 11)		
	2	10×13	M_{11}	(10, 8)	360	0.185
			M_8	(10, 2)	1583	0.215
				(10, 8)		
				(10, 12)		
			M_5	(8, 13)	1380.96	0.216
				(10, 4)		
				(10, 8)		
PCR streaming	1	8×15	M_{13}	(8, 3)	281	2.55
				(8, 3)		
			M_{12}	(8, 8)	281	2.55
				(8, 8)		
			M_{14}	(6, 3)	1680.09	4.1
				(6, 3)		
			M_{11}	(8, 13)	1184	4.1
				(8, 13)		
	2	10×17	M_{12}	(10, 8)	477	3.1
				(10, 8)		
			M_{14}	(6, 3)	1689	4.1
				(6, 3)		
			M_{11}	(10, 13)	645	2.55
				(10, 13)		

proposed in this paper is carried out as an offline step, it can also be extended for online error localization. This would make our method more effective in practice since significant time and resources can be saved by doing a timely online error detection and re-synthesis. We are currently working on this.

References

1. Ananthanarayanan, V., Thies, W.: Biocoder: a programming language for standardizing and automating biology protocols. J. Biol. Eng. **4**(1), 13 (2010)
2. Banerjee, A., Roychoudhury, A., Harlie, J.A., Liang, Z.: Golden implementation driven software debugging. In: Proceedings of the Eighteenth ACM SIGSOFT International Symposium on Foundations of Software Engineering, pp. 177–186. ACM (2010)
3. Bhattacharjee, S., Banerjee, A., Chakrabarty, K., Bhattacharya, B.B.: Correctness checking of bio-chemical protocol realizations on a digital microfluidic biochip. In: 2014 27th International Conference on VLSI Design and 2014 13th International Conference on Embedded Systems, pp. 504–509. IEEE (2014)
4. Bhattacharjee, S., Banerjee, A., Ho, T.Y., Chakrabarty, K., Bhattacharya, B.B.: On producing linear dilution gradient of a sample with a digital microfluidic biochip. In: 2013 International Symposium on Electronic System Design (ISED), pp. 77–81. IEEE (2013)
5. Chen, Y.H., Hsu, C.L., Tsai, L.C., Huang, T.W., Ho, T.Y.: A reliability-oriented placement algorithm for reconfigurable digital microfluidic biochips using 3-d deferred decision making technique. IEEE Trans. Comput.-Aided Des. Integr. Circ. Syst. **32**(8), 1151–1162 (2013)
6. Grissom, D., Brisk, P.: Path scheduling on digital microfluidic biochips. In: 2012 49th ACM/EDAC/IEEE Design Automation Conference (DAC), pp. 26–35. IEEE (2012)
7. Grissom, D.T., Brisk, P.: Fast online synthesis of digital microfluidic biochips. IEEE Trans. Comput.-Aided Des. Integr. Circ. Syst. **33**(3), 356–369 (2014)
8. Ho, T.Y., Zeng, J., Chakrabarty, K.: Digital microfluidic biochips: a vision for functional diversity and more than moore. In: Proceedings of the International Conference on Computer-Aided Design, pp. 578–585. IEEE Press (2010)
9. Hu, K., Bhattacharya, B.B., Chakrabarty, K.: Fault diagnosis for leakage and blockage defects in flow-based microfluidic biochips. IEEE Trans. Comput.-Aided Des. Integr. Circ. Syst. **35**(7), 1179–1191 (2016)
10. Hu, K., Yu, F., Ho, T.Y., Chakrabarty, K.: Testing of flow-based microfluidic biochips: fault modeling, test generation, and experimental demonstration. IEEE Trans. Comput.-Aided Des. Integr. Circ. Syst. **33**(10), 1463–1475 (2014)
11. Jaress, C., Brisk, P., Grissom, D.: Rapid online fault recovery for cyber-physical digital microfluidic biochips. In: 2015 IEEE 33rd VLSI Test Symposium (VTS), pp. 1–6. IEEE (2015)
12. Keszocze, O., Wille, R., Drechsler, R.: Exact routing for digital microfluidic biochips with temporary blockages. In: Proceedings of the 2014 IEEE/ACM International Conference on Computer-Aided Design, pp. 405–410. IEEE Press (2014)
13. Keszocze, O., Wille, R., Ho, T.Y., Drechsler, R.: Exact one-pass synthesis of digital microfluidic biochips. In: Proceedings of the 51st Annual Design Automation Conference, pp. 1–6. ACM (2014)
14. Luo, Y., Chakrabarty, K., Ho, T.Y.: Error recovery in cyberphysical digital microfluidic biochips. IEEE Trans. Comput.-Aided Des. Integr. Circ. Syst. **32**(1), 59–72 (2013)
15. Mazutis, L., Gilbert, J., Ung, W.L., Weitz, D.A., Griffiths, A.D., Heyman, J.A.: Single-cell analysis and sorting using droplet-based microfluidics. Nat. Protoc. **8**(5), 870 (2013)
16. Roy, P., Banerjee, A.: A new approach for root-causing attacks on digital microfluidic devices. In: AsianHOST, pp. 1–6 (2016)

17. Roy, S., Kumar, S., Chakrabarti, P.P., Bhattacharya, B.B., Chakrabarty, K.: Demand-driven mixture preparation and droplet streaming using digital microfluidic biochips. In: Proceedings of the 51st Annual Design Automation Conference, pp. 1–6. ACM (2014)
18. Sista, R., Hua, Z., Thwar, P., Sudarsan, A., Srinivasan, V., Eckhardt, A., Pollack, M., Pamula, V.: Development of a digital microfluidic platform for point of care testing. Lab Chip **8**(12), 2091–2104 (2008)
19. Su, F., Chakrabarty, K.: High-level synthesis of digital microfluidic biochips. ACM J. Emerg. Technol. Comput. Syst. (JETC) **3**(4), 1 (2008)
20. Su, F., Hwang, W., Chakrabarty, K.: Droplet routing in the synthesis of digital microfluidic biochips. In: Proceedings of Design, Automation and Test in Europe, 2006. DATE'06, vol. 1, pp. 1–6. IEEE (2006)
21. Thies, W., Urbanski, J.P., Thorsen, T., Amarasinghe, S.: Abstraction layers for scalable microfluidic biocomputing. Nat. Comput. **7**(2), 255–275 (2008)
22. Wu, P.H., Bai, S.Y., Ho, T.Y.: A topology-based eco routing methodology for mask cost minimization. In: 2014 19th Asia and South Pacific Design Automation Conference (ASP-DAC), pp. 507–512. IEEE (2014)
23. Xu, T., Chakrabarty, K.: Integrated droplet routing in the synthesis of microfluidic biochips. In: Proceedings of the 44th annual Design Automation Conference, pp. 948–953. ACM (2007)
24. Yeh, S.H., Chang, J.W., Huang, T.W., Yu, S.T., Ho, T.Y.: Voltage-Aware Chip-Level Design for Reliability-Driven Pin-Constrained Ewod Chips, vol. 33, pp. 1302–1315. IEEE (2014)
25. Zhao, Y., Xu, T., Chakrabarty, K.: Integrated control-path design and error recovery in the synthesis of digital microfluidic lab-on-chip. JETC **6**(3), 11 (2010)

Effect of Volumetric Split-Errors on Reactant-Concentration During Sample Preparation with Microfluidic Biochips

Sudip Poddar, Robert Wille, Hafizur Rahaman and Bhargab B. Bhattacharya

Abstract Recent microfluidic technologies offer suitable platforms for automating sample preparation on-chip, and typically on a digital microfluidic biochip, a sequence of (1 : 1) mix-split operations is performed on fluid droplets to achieve the target concentration factor of a sample. A (1 : 1) mixing model ideally mixes two unit-volume droplets followed by a (balanced) splitting into two unit-volume daughter-droplets. However, a major source of error in fluidic operations is due to unbalanced splitting, where two unequal-volume droplets are produced. Such volumetric split-errors occurring in different mix-split steps of the reaction path often cause a significant drift in the target-CF, the precision of which cannot be compromised in life-critical assays. In order to circumvent this problem, several error-recovery techniques have been proposed recently for DMFBs. Unfortunately, the impact of such fluidic errors on a target-CF and the dynamics of their behavior are not yet fully understood. In this work, we investigate the effect of multiple volumetric split-errors on various target-CFs during sample preparation. We also perform a detailed analysis of the worst-case scenario, i.e., when the error in a target-CF is maximized. This analysis may lead to the development of new techniques for error-tolerant sample preparation with DMFBs without using any sensing operation.

S. Poddar (✉) · B. B. Bhattacharya
Indian Statistical Institute, Kolkata, India
e-mail: sudippoddar2006@gmail.com

B. B. Bhattacharya
e-mail: bhargab.bhatta@gmail.com

H. Rahaman
Indian Institute of Engineering Science and Technology, Shibpur, India
e-mail: hafizur@vlsi.iiests.ac.in

R. Wille
Johannes Kepler University Linz, Linz, Austria
e-mail: robert.wille@jku.at

© Springer Nature Singapore Pte Ltd. 2020
R. Chaki et al. (eds.), *Advanced Computing and Systems for Security*,
Advances in Intelligent Systems and Computing 996,
https://doi.org/10.1007/978-981-13-8969-6_10

1   Introduction

A digital microfluidic biochip (DMFB) is capable of executing multiple tasks of biochemical laboratory protocols in an efficient manner. DMFBs support droplet-based operations on a single chip with high sensitivity and reconfigurability. Discrete volume (nanoliter/picoliter) droplets are manipulated on DMFBs through electrical actuation on an electrode array. Various fluid-handling operations such as dispensing, transport, mixing, split, and dilution can be performed on these tiny chips with higher speed and reliability.

Sample preparation imparts significant impact on accuracy, assay-completion time and cost, and plays a pivotal role in biomedical engineering and life science [1]. In the last few years, a large number of sample preparation algorithms had been developed [2–4]. Although droplet-based microfluidic biochips enable the integration of fluid-handling operations and outcome sensing on a single biochip, errors are likely to occur during fluidic operations due to various permanent or transient faults (e.g., unbalanced split due to imperfect actuation). For example, two daughter-droplets may be of different volume after split-operation while executing mix-split steps on a DMFB platform. Unbalanced split-errors, obviously pose a significant threat to sample preparation.

Albeit a number of cyber-physical-based approaches have been proposed for error-recovery [5–7], they do not provide any guarantee on the number of rollback iterations that are needed to rectify the error. Thus, most of the prior error-recovery approaches are non-deterministic in nature. On the other hand, the approach proposed in [8] performs error-correction in a deterministic sense; however, it assumes only the presence of single split-errors while classifying them as being *critical* or *non-critical*. In this paper, we focus especially on volumetric split-errors and investigate their effects on the target-CF during sample preparation. A detailed description of this analysis can be found in [9]. Based on our observations, a method for producing a target-CF within the allowable error-tolerance limit without using any sensor has been proposed [10].

The remainder of the paper is organized as follows. Section 2 introduces the basic principle of earlier error-recovery approaches. We describe the effect of one or more volumetric split-errors on the target-CF, in Sect. 3. Section 4 presents the worst-case scenario, i.e., when CF-error in the target-droplet becomes maximum. A justification behind the maximum CF-error is then reported in Sect. 5. Finally, we draw our conclusions in Sect. 6.

2   Error-Recovery Approaches: Prior Art

Earlier approaches attempt to recover the desired CF by re-executing a certain portion of an assay using pre-stored backup droplets [5], when an error is sensed. For example, the operations shown within the blue box in Fig. 1 are re-executed when

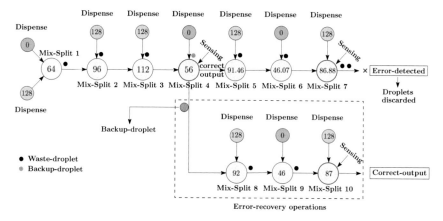

Fig. 1 Generation of a target-droplet by cyber-physical error-recovery approaches

an error is detected at the last checkpoint. However, such error-recovery mechanism suffers from significant overhead in terms of assay-completion time, reactant-cost, and uncertainties in termination due to randomness of split-errors.

3 Effect of Split-Errors on the Target Concentration

Generally, in the (1:1) mixing model (where two 1X-volume droplets are used for mixing operation), two 1X-volume daughter-droplets are produced after each mix-split operation. One of them is used in the subsequent mix-split operation and another one is discarded as waste droplet or stored for later use [2]. An erroneous mix-split operation may produce two unequal-volume droplets. Unless an elaborate sensing mechanism is used, it is not possible to predict which one of the resulting droplets (smaller/larger) is going to be used in the subsequent mix-split operation. Moreover, their effect on the target-CF becomes more complex when multiple volumetric split-errors occur in the mix-split path.

In order to analyze the effect of single volumetric split-error on the target-CF, we perform experiments with different erroneous droplets and present the results in this section. We assume an example target-$CF = \frac{87}{128}$ of accuracy level = 7. The mix-split sequence that needs to be performed using *twoWayMix* algorithm [2] for generating the target-CF is shown in Fig. 1. We consider the scenario of injecting 3% volumetric split-error at Mix-Split Step 4. Two unequal-volume daughter-droplets (a smaller and a larger) are produced after this step when a split-error occurs. The effect of the erroneous droplet on the target-CF depends on the choice of the daughter-droplet (smaller/larger) to be used next. For example, the effect of two errors on the target-CF (when the larger- or smaller-volume droplet is used at Mix-Split Step 4) is also shown in Fig. 2. The blue (green) box represents the scenario when the next operation

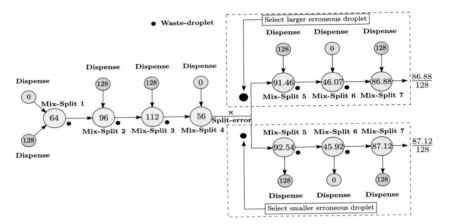

Fig. 2 Effect of choosing larger-/smaller-volume erroneous droplet on the target-$CF = \frac{87}{128}$

is executed with the larger (smaller) erroneous droplet. It has been seen from Fig. 2 that the CF-error in the target increases when the smaller erroneous droplet is used in the mixing path compared to the use of the larger one. We also observe that the CF-error in the target-CF increases when the magnitude of volumetric split-error increases.

In order to find the effect of multiple volumetric split-errors on the target-CF, we perform several experiments. We continue with the example target-$CF = \frac{87}{128}$ of accuracy level = 7, and inject 7% volumetric split-error simultaneously at different mix-split steps of the mixing path. During simulation, we assume that the larger erroneous droplet is always used later when a split-error occurs in the mix-split path (i.e., ϵ is positive). It has been observed that CF-error in the target-droplet rapidly grows to $\frac{0.08}{128}$ and $\frac{0.17}{128}$ when two or three such split-errors are injected in the mix-split path.

4   Worst-Case Error in the Target-CF

So far, we have analyzed the effect of multiple volumetric split-errors on a target-CF when a larger erroneous droplet is selected following each mix-split step. However, in a "sensor-free" environment, multiple volumetric split-errors may consist of an arbitrary combination of large and small daughter-droplets. Hence, further analysis is required to reveal the role of such random occurrence of volumetric split-errors and their effects on the target-CF.

In order to facilitate the analysis, we define "error-vector" as follows: An error-vector of length k denotes the sequence of larger or smaller erroneous droplets, which are chosen corresponding to k mix-split-errors in the mixing path. For example, an error-vector $[+,\phi,-,\phi,\phi,+]$ denotes volumetric split-error in Mix-Split Step 1, Step

3, and Step 6, where ϕ denotes no-error. In Step 1, the larger droplet is passed to the next step, whereas in Step 3, the smaller one is used in the next step, and so on. For k volumetric split-errors, 3^k error-vectors are possible.

We perform simulated experiments for finding the effect of different error-vectors (# error-vectors = 64) for the target-$CF = \frac{87}{128}$ and report the generated CFs in Table 1 for some error-vectors. It has been observed that the CF-error exceeds allowable error-tolerance limit in all such cases. Based on exhaustive simulation, we observe that the CF-error in the target-CFs becomes maximum (1.977) for the error-vector $[-,+,+,-,+,-]$ (at the 57th position on the X-axis in Fig. 3) for the target-$CF = \frac{41}{128}$ and $\frac{87}{128}$ (complement of $\frac{41}{128}$).

Table 1 Effect of some error-vectors of length 6 on the target-$CF = \frac{87}{128}$ for split-error = 7%

Error-vector	Produced $CF \times 128^a$	Produced CF-error$\times 128$	CF-error$\times 128 < 0.5$?
$[+,+,+,+,+,+]$	85.58	1.42	No
$[+,-,+,+,+,+]$	85.53	1.47	No
$[+,-,-,+,+,+]$	85.26	1.74	No
$[+,-,+,+,-,+]$	85.08	1.92	No
$[-,+,-,-,+,-]$	88.78	1.78	No
$[-,+,+,-,-,-]$	88.82	1.82	No
$[-,+,-,-,-,-]$	88.64	1.64	No
$[-,-,-,-,-,-]$	88.61	1.61	No

^{a}Results are shown up to two decimal places

Fig. 3 Value of (CF-error$\times 128$) for all possible error-vectors with 7% split-error for the target-$CF = \frac{41}{128}$ and $\frac{87}{128}$

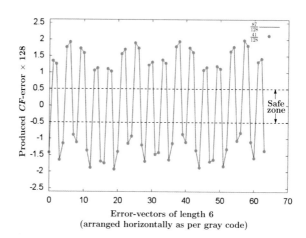

Error-vectors of length 6
(arranged horizontally as per gray code)

5   Maximum CF-Error: A Justification

We have performed analysis and further experiments to study the properties of *CF*-error in a target-*CF*. These results reveal how the problem of error-tolerance can be handled in a more concrete fashion. Consider a particular target-$CF = C_t$ and its dilution tree. Let the current mix-split step be i (other than the last step, where the occurrence of split-error does not matter), and the intermediate-*CF* arriving at i be C_i. If a 1X sample (buffer) droplet is added in this step, it produces $CF = \frac{C_i+1}{2}$ (= $\frac{C_i}{2}$), assuming that the volume of the droplet arriving at i is correct (1X). Consider the first case, and assume that the droplet arriving at i suffers a volumetric split-error of magnitude ϵ at the previous step. Hence, after mixing with a sample droplet, the intermediate-*CF* will become: $\frac{C_i(1+\epsilon)+1}{2+\epsilon}$; the sign of ϵ is set to positive (negative) when the incoming intermediate-droplet is larger (smaller) than the ideal volume 1X. Thus, the error (E_r) in the intermediate-*CF* becomes:

$$E_r = \frac{C_i+1}{2} - \frac{C_i(1+\epsilon)+1}{2+\epsilon} = \frac{\epsilon(1-C_i)}{4+2\epsilon} \tag{1}$$

From Eq. 1, it can be observed that the *CF*-error will be more if a droplet of smaller-volume arrives at Step i compared to the case when a larger-volume droplet arrives at the mixer. We perform an experiment assuming $\epsilon = +0.07$ or -0.07 in one mix-split step, for all values of intermediate-*CF*s and observed that a negative split-error always produces larger *CF*-error in the target-*CF* for a single split-error (error-vector of length 1). We also perform simulation by varying C_i from 0 to 1, and ϵ from -0.07 to 0.07 in Eq. 1 and report the calculated *CF*-errors as three-dimensional (3D) plot in Fig. 4. We observe that simulation results favorably match with theoretical results, i.e., the negative split-error (single) always produces larger *CF*-error for a single split-error. However, the error-expression becomes increasingly

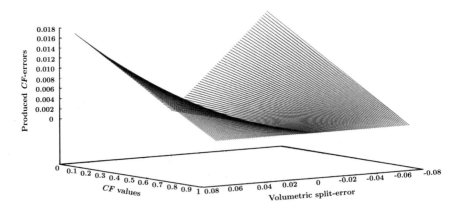

Fig. 4 *CF*-error at the next mix-split step (for positive and negative single split-error)

complex for multiple split-errors. For example, we perform experiments to study the fluctuations of the error in a particular target-CF for all combinations of error-vectors and observe several peaks up and down as shown in Fig. 3. From the above analysis and experimental results, we conclude that it is hard to formulate a mechanism that will identify the exact "maximum-error-vector" without doing exhaustive simulation.

6   Conclusion

In this paper, initially, we analyze the effect of single volumetric split-errors (with larger- or smaller-volume erroneous droplet) on the target-CF and observed that (both theoretically and experimentally) the maximum value of the CF-error in the target-droplet occurs for a negative split-error. We also observe that the CF-error in a target-droplet increases with increasing magnitude of the split-error. Next, we perform various experiments to observe the effect of multiple CF-errors on the target-CF and notice that it may be affected by any combination of erroneous droplets (smaller/larger) during the execution of mix-split operations. We also observe that the CF-error in a target-droplet increases when the target-CF is affected by a large number of split-errors. We perform rigorous analysis to identify the error-vector that causes the maximum CF-error in the target-droplet. Although it is difficult to identify an error vector that maximizes the CF-error in the target for multiple split-errors without doing exhaustive simulation, full-scale error cancelation can be achieved by choosing an appropriate initial error-vector and redesigning the reaction paths of the dilution assay as described elsewhere [10].

References

1. Srinivasan, V., et al.: An integrated digital microfluidic lab-on-a-chip for clinical diagnostics on human physiological fluids. Lab Chip **4**, 310–315 (2004)
2. Thies, W., et al.: Abstraction layers for scalable microfluidic biocomputing. Nat. Comput. **7**, 255–275 (2008)
3. Poddar, S., et al.: Optimization of multi-target sample preparation on-demand with digital microfluidic biochips. IEEE TCAD **38**, 253–266 (2019)
4. Bhattacharjee, S., et al.: Dilution and mixing algorithms for flow-based microfluidic biochips. IEEE TCAD **36**, 614–627 (2017)
5. Luo, Y., et al.: Error recovery in cyberphysical digital microfluidic biochips. IEEE TCAD **32**, 59–72 (2013)
6. Luo, Y., et al.: Real-time error recovery in cyberphysical digital-microfluidic biochips using a compact dictionary. IEEE TCAD **32**, 1839–1852 (2013)
7. Luo, Y., et al.: Biochemistry synthesis on a cyberphysical digital microfluidics platform under completion-time uncertainties in fluidic operations. IEEE TCAD **33**, 903–916 (2014)
8. Poddar, S., et al.: Error-correcting sample preparation with cyberphysical digital microfluidic lab-on-chip. In: ACM TODAES, vol. 22, pp. 2:1–2:29 (2016)
9. Poddar, S., et al.: Dilution with Digital Microfluidic Biochips: How Unbalanced Splits Corrupt Target-Concentration, CoRR (2019). arXiv:1901.00353
10. Poddar, S., et al.: Error-oblivious sample preparation with digital microfluidic lab-on-chip. IEEE TCAD (2018). https://doi.org/10.1109/TCAD.2018.2864263

Author Index

© Springer Nature Singapore Pte Ltd. 2020
R. Chaki et al. (eds.), *Advanced Computing and Systems for Security*,
Advances in Intelligent Systems and Computing 996,
https://doi.org/10.1007/978-981-13-8969-6

Printed in the United States
By Bookmasters